Ludwig Büchner, Annie Wood Besant

Mind in animals

Ludwig Büchner, Annie Wood Besant

Mind in animals

ISBN/EAN: 9783337228552

Printed in Europe, USA, Canada, Australia, Japan

Cover: Foto ©berggeist007 / pixelio.de

More available books at **www.hansebooks.com**

MIND IN ANIMALS.

BY

PROFESSOR LUDWIG BÜCHNER,

AUTHOR OF "FORCE AND MATTER," "MAN, AND HIS PLACE IN NATURE," ETC., ETC.

Translated, with the Author's permission, from the German of the Third Revised Edition,

BY

ANNIE BESANT.

LONDON:
FREETHOUGHT PUBLISHING COMPANY,
28, STONECUTTER STREET, E.C.

ADVERTISEMENT.

In issuing, under the name of the "INTERNATIONAL LIBRARY OF SCIENCE AND FREETHOUGHT," a series of books of which the present volume is the first, the FREETHOUGHT PUBLISHING COMPANY desires to place at the service of English Freethought the weapons wielded against superstition in foreign countries as well as those forged in England itself. The writings of foreign scientists are not as well known in England as their merit deserves; there are some valuable text-books—such as those of Gegenbaur and of Thomé—which have their place on the bookshelf of the student; but the aim of the FREETHOUGHT PUBLISHING COMPANY is to issue such works as will reach the general reader, as well as the scientific student, and render Büchner, Häckel and others as well-known to the English public as are Huxley and Darwin. German science is one of the glories of the world; it is time that it should lend in England that same aid to Freethought which in Germany has made every educated man a Freethinker. France will contribute to this new library some of the works of her leading sceptics; and Italy, also, has furnished help to Freethought which will not be forgotten. English and American works will not be excluded, and it is hoped that a real service will be done to progress by thus popularising in one country the knowledge gained in many lands by the earnest searchers after truth.

PREFACE TO THE FIRST EDITION.

IF the author of this book had had to choose a title for it after it was written he would have called it "A Romance of the Animal World." For the contents both seem and are romantic and wonderful, although—with the exception of such doubtful or possibly doubtful observations as are given on the authority of the writer himself—there is nothing therein which does not rest on scientific investigation or on the evidence of trustworthy observers, who, at different times and in different places, have had co-incident experiences, and whose accounts bear the stamp of sober research, and are the simple description of things they actually saw. But as histories (of nations as of individuals), related just as they really happened and still are daily happening, are full of more wonderful and more startling occurrences, of grander tragedy and more irresistible comedy, of more apparently impossible and incredible things and events than are told in fiction, and leave far behind them the boldest fancies of poets and novelists, so it is also with nature; she is wont, the more we peer into her secrets, to bring the most marvellous, the mightiest and the most astonishing forms out of the simplest and the least differentiated. That "Mind in Animals" especially is in reality a far other, higher and more complex thing than had hitherto been generally conceived, and indeed than the ruling schools of philosophy desired (and still desire) to admit, can be unknown to none who is acquainted with animals, not alone from hearsay and from philosophic writings, but from his own intercourse with them, from his own observation, or from the works and teachings of real and unprejudiced observers. For such observation furnishes continually, and with overwhelming fullness, the most startling and incontrovertible examples and proofs, that between the thinking, willing, and feeling of men and of animals there is the most striking similarity, and often a mere difference of degree. But even among comparatively educated people it has been little thought and felt that this rule applies also to those classes of animals which appear to be so far below us as those treated of in the present work; our intellectual vanity will have to submit to bitter humiliation and rebuke in contemplating the proceedings—or the societies and deeds—of these unjustly despised, but yet, in spite of their minuteness, wonderful creatures. But the greater the humiliation from the one point of view, the greater from the other is the satisfaction arising from the renewed proof of the sublime unity of Nature; and hence that the same intellectual or spiritual principle, call it reason, understanding, soul, instinct, or propensity, pervades the whole organised series, even if in the most manifold modifications and variations, from below to above, from above to below.

Starting from this last standpoint, the author has not thought it necessary to widen the circle of his observations over the whole of the comparatively narrow and yet infinitely wide and rich sphere of intelligent insect life; he considers it better, according to the true and ancient proverb, *Multum non multa*, to treat a single species thoroughly, rather than many species cursorily and superficially, thus falling into the common

PREFACE TO THE FIRST EDITION.

IF the author of this book had had to choose a title for it after it was written he would have called it "A Romance of the Animal World." For the contents both seem and are romantic and wonderful, although—with the exception of such doubtful or possibly doubtful observations as are given on the authority of the writer himself—there is nothing therein which does not rest on scientific investigation or on the evidence of trustworthy observers, who, at different times and in different places, have had co-incident experiences, and whose accounts bear the stamp of sober research, and are the simple description of things they actually saw. But as histories (of nations as of individuals), related just as they really happened and still are daily happening, are full of more wonderful and more startling occurrences, of grander tragedy and more irresistible comedy, of more apparently impossible and incredible things and events than are told in fiction, and leave far behind them the boldest fancies of poets and novelists, so it is also with nature; she is wont, the more we peer into her secrets, to bring the most marvellous, the mightiest and the most astonishing forms out of the simplest and the least differentiated. That "Mind in Animals" especially is in reality a far other, higher and more complex thing than had hitherto been generally conceived, and indeed than the ruling schools of philosophy desired (and still desire) to admit, can be unknown to none who is acquainted with animals, not alone from hearsay and from philosophic writings, but from his own intercourse with them, from his own observation, or from the works and teachings of real and unprejudiced observers. For such observation furnishes continually, and with overwhelming fullness, the most startling and incontrovertible examples and proofs, that between the thinking, willing, and feeling of men and of animals there is the most striking similarity, and often a mere difference of degree. But even among comparatively educated people it has been little thought and felt that this rule applies also to those classes of animals which appear to be so far below us as those treated of in the present work ; our intellectual vanity will have to submit to bitter humiliation and rebuke in contemplating the proceedings—or the societies and deeds—of these unjustly despised, but yet, in spite of their minuteness, wonderful creatures. But the greater the humiliation from the one point of view, the greater from the other is the satisfaction arising from the renewed proof of the sublime unity of Nature; and hence that the same intellectual or spiritual principle, call it reason, understanding, soul, instinct, or propensity, pervades the whole organised series, even if in the most manifold modifications and variations, from below to above, from above to below.

Starting from this last standpoint, the author has not thought it necessary to widen the circle of his observations over the whole of the comparatively narrow and yet infinitely wide and rich sphere of intelligent insect life; he considers it better, according to the true and ancient proverb, *Multum non multa*, to treat a single species thoroughly, rather than many species cursorily and superficially, thus falling into the common blunder of writers on Animal Psychology, who are wont rather to be dazzled by the overwhelming mass of materials than to be enlightened thereby. For it is just in the individual and in the minute, rather than in the general, that the truth of the above-named principle shines out most clearly and most strikingly; and it is here, at the same time, that we shall find the most easily recognisable land-marks for further researches in this direction. And this is so, although the proposed sketch does not pretend to lay claim to completeness, and although the writer has been compelled by the limits of a popular work to confine himself to the most necessary and to the most commonly known things, and to make many distasteful abridgements. Even those who turn away from the philosophical meaning and tendency of the observations herein recorded, and only desire entertainment, or entertaining instruction, will also, the author hopes, not be disappointed in reading the book, although the thoughtful reader is enabled to enjoy a special pleasure of his own in the likeness to the doings of men which lies on the surface and indeed presses itself on his attention. The well-known philosopher, Daumer, who has passed from Radicalism to piety, has indeed made the characteristic remark that many glimpses into the minds of animals "must make one shudder." But this can only be when anyone clings to the antiquated notion that animals are beings entirely and essentially distinct from men, and that all they do can only be the outcome of unconscious and unchangeable instincts. All others must feel a true intellectual joy when they recognise in the psychical world that same law of the origin and development of organic life, that has been demonstrated in physical things by Lamarck, Oken, Darwin, Häckel, and others

It will naturally be understood that the narrowing of the subject to a relatively small field of Animal Psychology compels the author to use only a very small number out of the many hundreds of facts and observations on the intellectual life of animals, which have been sent to him from all parts of the world in answer to his public request, and for which he here again returns his public acknowledgment; further, the greater number of these communications give the result, as might be expected, of daily and private observations on the more accessible animals. The author therefore ventures to refer his respected correspondents to a later work from his own pen which, as compared with the present book, undertakes a far wider task. He will therein endeavor, by means of a psychological classification, to trace the different affections and manifestations of man's emotional and intellectual life throughout the great circles of the animal world. Here also most of the communications will be found under the names of the several observers, and the author trusts that this will be accepted as his personal acknowledgment for individual help.

Darmstadt, October, 1876.

PREFACE TO THE THIRD EDITION.

THE great interest which has been aroused by the publication of the improved edition of this work makes it possible to issue this third edition, despite the comparatively short time that has elapsed and the unfavorable nature of the times for literary enterprises. The book has again been most carefully revised, and has been enriched and rendered more complete by many new additional facts taken from the researches and communications of Lubbock, McCook, Graber, Espinas, Taschenberg, Müller, Dzierzon, and others. Especially worthy of thanks are the researches carried on in Philadelphia by the Rev. H. C. McCook, on the habits of ants and spiders; these, like the communications of the above named investigators, confirm and complete in almost every particular the facts already published; this is notably the case with the results given in the admirable work of Dr. A. Forel on the ants of Switzerland. We cannot deny ourselves the pleasure of here quoting the excellent remarks of M. A. Espinas, the author of the book on "Animal Societies" (translated into German by Vieweg, 1879), although we do not thereby wish to give the idea that our own volume owes its existence to any motives save to those of simple love of truth and of popular instruction or instructive amusement. M. Espinas says: "To elevate animal societies is at the same time to elevate those of men, who so widely surpass and rule them. We think that we shall better promote civilisation if we show that man is the last step of an advancing progress, than if we isolate him in the universe and leave him to rule a world bare of intelligence and of emotion." Equally well does the same author speak in another place, wherein he argues against those who above all things will not admit any comparison between the intellectual powers of men and beasts: "It is clear that we can only comprehend an intelligence, of whatever kind it may be, when we can find its analogue in our own intellectual life. This is a necessity of Animal Psychology to which we must absolutely hold fast."

The promises made by the author at the close of the preface of the first edition have been partly redeemed by the issue, by M. A. Hofman, of Berlin, of the work on "Love and love habits of the Animal World," as one of the publications of the "Universal Society of German Literature." The author further hopes that he may have time and opportunity to fulfil the remaining part of his promise before very long. It will doubtless interest readers of this volume to hear that a Dutch translation has appeared under the title of "Uit het Leven der Dieren, vertaald door R. E. de Haan" (Nimwegen, published by Bloomert and Timmerman), and that a French translation of this third edition will shortly be issued in Paris by Reinwald.

Darmstadt, March, 1880.

INTRODUCTION.

CHAPTER I.

HISTORICAL REVIEW.

THE question of mind in animals and of their intellectual capacities as compared with those of men is as old as man's thought; it can scarcely be accepted as a brilliant testimony to human philosophy and its progress, that the different points of view from which this question has been judged stand out against each other to-day with almost the same distinctness as was the case some thousand years ago, although lately the influence of the Darwinian theory, and the more accurate knowledge of the remarkable facts of heredity, have thrown a heavy weight into the scale of the opinion hitherto rejected by the majority. This opinion has been urged or denied less from scientific than from egoistical motives; it was feared lest man and his place in nature should be lowered and degraded if animals were allowed the possession of intellectual powers like or allied to those of man. Just as if "our superiority over the animals" (as Lord Brougham says in his " Discourse on Instinct") "was not great enough to banish and make ridiculous every feeling of jealousy in this respect, even if we regard the difference between ourselves and them as a question of degree and not of kind."

There was indeed in their exceedingly slight knowledge of animals and of their habits an excuse for the philosophers of antiquity, which cannot be admitted for the philosophers of to-day. Nevertheless, Anaxagoras, with philosophic insight, calls man the wisest of animals; Socrates calls him a beautiful, and Plato a civilised animal. Their disciple,

Aristotle—who far surpassed his predecessors in scientific knowledge—approached more nearly to the solution of the question, for he had caught a glimpse of the gradations of organised beings. He sees in the minds of animals traces of the properties of human minds and human reflection, and maintains that the mind of the child scarcely differs in aught from the mind of the animal ("Natural History," Book 8). He regards the elephant as the most intelligent of animals. The Roman Pliny, although too credulous, does the same in relating wonderful anecdotes about animals. Also the Roman poet, Virgil (70 B.C.), speaks very lovingly in his poems about the breeding of animals, and in describing the wonderful doings of bees declares that a portion of the divine spirit dwells in these creatures. Plutarch (B.C. 50) in his treatise on reason in animals makes himself merry over the opinion—taught in the schools of the cynics and stoics and still defended to-day— that animals in reality possess neither emotion nor thought, and that the identity of their actions with those of men is only apparent. " As for those," he says, " who judge so clumsily and are so barefaced as to maintain that animals feel neither joy, nor anger, nor fear, that the swallow has no forethought, and the bee no memory, but that it is a mere appearance when the swallow shows forethought, or the lion anger, or the hind timidity—I do not know how they would answer those who should say that they must then also admit that animals do not see, nor hear, nor have voices, but that they only apparently see, hear, and have voices ; that in fact they do not really live at all, but only appear to have life. For the one contention would not be more antagonistic to manifest fact than is the other."

Plutarch seems also to embrace the opinion, about which there is now so much controversy, that the difference between animals of the same race is not nearly so great as that between man and man.

The great Roman physician, Claudius Galen, of Pergamus, whose system of medicine ruled the world for more than a thousand years, gives it plainly to be understood in his writings that he ascribes reflection and power of determination to animals, and that they only differ from men as to degree. He also, like Anaxagoras, calls man the wisest of animals.

The first writer of the Christian era who troubled himself about animals, and combatted their more and more strongly emphasised inferiority to man, was Celsus, who lived in the second century after Christ, and who followed the materialistic philosophy of the Epicureans as adapted by the Platonists. He fought with wit and acuteness against Christianity, and also against the Judæo-Christian theory that everything was created for the sake of man, and that he was the final cause of the universe. He maintained, as regards animals, that their bodies differed in no important respect from those of men, and that in intellectual qualities they were in many things higher rather than lower than men, since they had a kind of intelligible government, and observed justice and love. His proofs in support of this argument he draws from the life of bees and ants—and with what justice the reader of this book will find abundant evidence.

" If men," proceeds Celsus, " want to separate themselves from animals because they inhabit towns, make laws, and set up a government, yet all this proves nothing; bees and ants do the same. Bees have their king, whom they accompany and obey; they have their wars, their victories, their massacres of the conquered; they have towns and suburbs, regular hours of work, penalties for the lazy and the bad; they hunt and punish hornets . . . ." He awards the same praise to the ants and to their prudence and care for the future. They help each other to carry heavy loads. " Out of the seeds and fruits which they collect they put on one side those which have begun to sprout, so that they may not affect the others, and that they may serve as food for the winter." They speak to each other when they meet, and do not mistake their road. Celsus even thinks that they have their own burying grounds. " If anyone were able to look down upon the earth from heaven, what difference would he see between the works of men and those of ants and of bees?"

The Christian Middle Ages, enemy of all natural investigation, could evidently make no peace with such theories. In spite of the vigorous opposition of Rorarius, the learned nuncio of Clement VII. to the Court of the Emperor Ferdinand in Hungary, who brought forward a mass of facts in support of the reasoning powers of animals, and maintained, like Celsus, that they often put their reason to a better use than did men, the Church upheld the contrary view, going even as far as the famous, or at least notorious, contention of the French philosopher, Descartes (1596 to 1650), which, as is well-known, takes away from animals all conscious feeling and emotion, and only regards them as living machines, or as automata. Descartes, however, is not the only holder of this opinion. He borrowed it from a predecessor, the Spanish physician, Gomez Pereira, who in his "Antoniana Margarita," published in the sixteenth century, first maintained that animals had neither intellectual feeling nor capacity of thought, and that, above all, they had no minds, but were only machines controlled by external circumstances. Descartes, whose whole philosophy rests on the dualism of matter and spirit, admits nevertheless that animals do many things better than men ; but they therein follow, he asserts, only a blind instinct, or a mechanical impulse communicated through their external organs, just as a watch, an artificial machine, measures time better than a man, with all his intellect and reason. According to Descartes, the feelings and emotions of animals are an empty show ; a welcome piece of news for animal-tormentors! "After the error of Atheism," says Descartes, "there is none which leads weak minds further from the path of virtue than the idea that the minds of animals resemble our own, and therefore that we have no greater right to a future life than have gnats and ants, while, on the contrary, our mind is quite independent of the body, and does not therefore necessarily perish with it."

This extreme opinion made a great success in its time, so that no man could call himself a Cartesian without declaring that animals were machines.

Besides, the all-powerful devil of the Middle Ages got mixed up in the matter, and was held to be the author of the unmistakable manifestations of reason in animals, by those who sought for some ground for them ; while, on the contrary, others did not hesitate to impute this same authorship to the Almighty Creator of heaven and earth, through the mediation of so-called instinct, which God had implanted in the minds of animals for their preservation and increase—a guiding and irresistible natural propensity, inborn, unchangeable, independent of experience and training, and acting appropriately without consciousness of the object aimed at. The word instinct is derived from the Latin *instinguere*, to excite or to allure, and therefore necessarily connotes an exciter or allurer. Therefore Cæsalpinius says, very properly, from this point of view: *Deus est anima brutorum*—God is the mind of animals.

It is evident that under such circumstances the question of the mind of animals could never become a scientific study; scientific investigation of the intellectual and emotional powers of animals in comparison with those of man, in fact comparative psychology, was impossible. The subject was either dealt with as a mere collection of curiosities, an innocent amusement, an intellectual pastime, or, more usually, the purely theological standpoint was taken, and the endeavor was made to use this theme—as so many similar ones drawn from Nature—merely as an object of pious wonder.

The Cartesian theory had given rise to much controversy in its time, and the hostile philosophy of Leibnitz, teaching organic gradation and the unbroken unity of all living things, both of men and of animals, had called up against it a host of writers. The most important of these productions for our position was the small work of Jenkin Thomasuis, published in 1713, who supported against Descartes—and in the spirit of his own time and of Leibnitz—the immateriality, and therefore the immortality, of the animal soul. The German editor of the book, Professor Bajer, also declared against instinct, and held that among the various opinions of the learned on the mind of animals, that which saw in animals the analogue of the human mind was the most conformable to man's natural judgment, and the most useful for throwing light on animal actions. In like fashion Professor Reclam, another author on the same question ("Body and Mind," 1859, p. 384), wrote very well: "We therefore conclude that we ought entirely to give up the expression 'instinct,' for we only can and should apply it to such actions of animals as we can explain in no other way, and we ought, remembering Kepler's warning, to seek for every other explanation before we use a word so indefinite and so apt to mislead." In fact, those who deny this, and who will not compare the intellectual faculties of animals with those of men, must renounce all scientific

B

conceptions of these faculties, since there is no other rule wherewith to measure them, and the word "instinct"—as will presently be further shown—is only a paraphrase of our ignorance, and depends in countless cases on demonstrably false representations.* The French philosopher, Condillac, the able tutor of the Infant of Parma, who, by his victorious struggle against the innate ideas of Locke, gave the death-stroke to the vanishing relics of the Cartesian philosophy, had used the argument against Descartes that animals were far removed from machines, since they felt like ourselves, avoided danger, gained expertness, and supplied their own wants as did human beings. "Man is wont to say," remarks Condillac, "that animals obey instinct and man reason, without knowing what is to be conveyed by these two words. The actions of animals can only be explained on three principles; either as the result of mere mechanism, or of blind impulse which neither reasons nor judges, or as the outcome of something which reasons, judges and understands. Since I have proved that both the first explanations are utterly unsatisfactory, the last alone remains." Linnæus, Buffon—who made the admirable remark that we were compelled to marvel the more at the intelligence of animals, the more we observed and the less we theorised—Voltaire, G. F. Meier (in his famous "Search after a new system of Animal Intelligence," 1750), C. Bonnet, and many others spoke more or less against the Cartesian philosophy. The last especially, an excellent naturalist and a distinguished thinker (1770), refers to the contrivances of insects, especially of wasps and bees, and to the artistic talent of the beavers, which last are brought in immediately after the bees. (!)

Even the Jesuit father, Bonjeant, who found so much intelligence in animals that he thought it could only be due to the help of the devil or devils, turned against Descartes with the words: "All the Cartesians in the world will never persuade me that a dog is a mere machine. Imagine a man who should love his clock as a man loves his dog, and who should pet it because he believed it loved him and was of opinion that it struck the hours consciously and out

---

* Compare on this point the admirable treatment of the mind of animals, by L. H. Morgan, in "The American Beaver and his Works," p. 148, etc.

of friendship for him. Yet, if Descartes be right, that is exactly the absurdity committed by all those who believe that their dog is faithful to them and loves them. I see how my dog runs to me when I call him, caresses me when I coax him, trembles and runs away when I threaten him, obeys when I order him, and how he exhibits all the outward signs of the distinct emotions of joy, grief, pain, fear, desire, love, and hate.* And if all the philosophers in the world should come and try to convince me, I should never be able to persuade myself that an animal is a machine; this feeling will always set men against the philosophy of Descartes."

The great fable-teller, La Fontaine, has made fun of the Cartesian theory of animal-machines in several admirable poems (the fables of the two rats, the fox and the egg).

Descartes, however, found his most distinguished opponent in the French Inspector of Forests, G. Leroy, who, in consequence of his office in the royal gardens and forests at Marly and Versailles, had good opportunities of observing dogs and wood-animals. In the time of Leroy and Buffon, the question of the intelligence of animals had been for a hundred years and more a question of pure philosophy and speculation; it only entered into the circle of experimental observation with these two investigators. The first letters of Leroy—who was one of the famous French Encyclopædists—on the intelligence and perfectibility of animals, were published in 1764, under the name of a "Nuremberg Physician;" Leroy at that time would have been in danger of a prosecution by the Sorbonne, had he sought to prove that animals were not mere machines, but that they possessed all the marks of reason and improveability, as well as of feeling, thought and prudence. Want and need, fear and danger, and so on, are, with him, the guiding springs of the intellectual development of animals, which, as for example the wolf, communicate with each other, often hunt in packs, discover snares and utilise experience. With use, says Leroy, reason grows in animals, and the capacity of employing their minds; there is a very great difference between a young and an old wolf or fox. The young animals, partly from lack of reflection and experience, partly from overmastering timidity, commit a number of mistakes which the old ones avoid. The dog especially, owing to the chase and to his contact with man, learns to an extraordinary extent, and thinks out particular tricks for trapping game. According to Leroy, animals must have a common tongue, however little we may understand of it; it is impossible that their various communications can be made without speech, and they possess all that is supposed to be necessary to speech, such as faculties of thought, comparison, judgment, decision, reflection, etc. Leroy had then more accurate views about an animal-language than our great language-investigator, Max Müller, who calls speech the Rubicon between men and animals which will never be crossed.*

That which must most interest and astonish us in Leroy, however, who quite ignores instinct and refers everything to reason, is that he has already grasped the power and significance of the transmission of qualities acquired during life, and that he makes the important and fruitful remark: "that all that we regarded as mere blind mechanism in animals, was perhaps the result of long-since acquired habits which had been handed down from generation to generation." P. Floureus ("De l'Instinct et de l'Intelligence des Apimaux, 5 ed., p. 41) calls the investigations of Leroy the deepest which had until then been made into the intellectual capabilities of animals. After Leroy, it was the great naturalist, F. Cuvier, who, by means of observations made on a young ourang-outang kept by him in captivity, was led to declare that an animal was able "to combine various ideas in its head and to draw a conclusion from them;" and this

---

* The voice organs of men and animals are similar in construction; there is neither anatomical nor physiological distinction between them. Indeed, this organ in some animals, as in birds, possesses a flexibility and an imitative power exceeding those of men. Animals communicate with each other by means of tones and gestures, as do men; in both speech proceeds from imitation and hearing. Hence deaf people are also stupid. If speech, as many philosophers maintain, were innate in man, then men would speak even without hearing. The wild amongst animals, and children growing up alone amongst men, do not speak, but only utter cries, which are often almost the same in men and animals, as for instance the cries of astonishment or of fear. Also the language of primitive man has been much enriched by the imitation of natural sounds, education through the ear (Onomatopœia). Parrots can be trained to talk just as children are. (See D. S. Wilks, in the "Journal of Mental Science.") On the natural development or origin of speech see further the author's "Man and his place in Nature."

although the unprejudiced sensational philosopher, Locke, had doubted the capacity of animals for abstract thought, and had laid just in this the distinction between animal and human minds.

In spite of all this the old strife, whether animals were machines or thinking and conscious creatures, went on as vigorously and as indecisively as ever, and found great support in the ignorance of the laity, or the mass of the people, who even down to the present day are more inclined to embrace the Cartesian ideas than those opposed to them. The great and famed era of philosophic speculation at the end of the last and the commencement of the present century mastered very little of this difficult question, owing to its fondness of theoretic, and its dislike of experimental methods. Even the great Königsberg sage, by means of whose learning many now strive so vainly to rebuild the crumbled philosophy of the schools, stood helpless and powerless before it, owing to his philosophical prejudices, as before the question of the relations between the brain and the soul, or the brain and the mind.* For him the animal, like the plant and the mineral, is a mere thing, and is quite excluded from right and morality, which belong only to men. It has no reason, no judgment, it knows no rights and no duties, and is not capable of education but can only be disciplined. Man only owes to animals the duty of kindness, and this not for the animals' sake but for his own. "And such things were taught," cries Sheitlin, indignantly, "in the Kantian primer of morality and orthodoxy fifty years ago."

From similar views about animals started Kant's famous disciple, the philosophical idealist and metaphysical egoist, Fichte, who, from the standpoint of so-called "pure reason," declares that animals are things without freedom, personality, reason or rights. O Philosophy, noblest and highest of all sciences, how pitiable thou appearest in the eyes of the lovers of Truth, when thou submittest to be led, not by experience and fact, but by the clinging to preconceived opinions, and by philosophic rules and axioms accepted as eternally valid!

Far better than Kant and Fichte did their noble and

---

* Compare the author's treatment of the brain in the second volume of his "Physiological Types." (Leipsig, Thomas, 1875.)

talented contemporary, Herder. grasp the nature of animals, whom, in his "Thoughts for a Philosophy of Human History," he calls with keen insight "the elder brethren of men." Brain development and erect position have made men into men; yet in the animal-kingdom we see the rudiments of all the higher moral and intellectual faculties of man, such as reason, speech, art, freedom, etc.

With this last declaration, Herder approaches very nearly the standpoint of modern thought, which no longer recognises in animals a difference of kind, but only a difference of degree, and which sees the principle of intelligence developing through one endless and unbroken series, gradually and slowly, by way of countless acquisitions, inheritances, and transmissions, from its lowest stage to its highest. "The principle exists without doubt," says the anti-materialist, Agassiz, on this point ("Treatise on Natural History," U.S.A.), "and whether it be called soul, reason, or instinct, it displays in the whole series of organised beings a succession of closely interwoven phænomena." According to Huxley, the admirable English naturalist ("Natural History Review, 1861"), no impartial judge can doubt that the roots of all those great capabilities, which give to man his immeasurable superiority over all other living things, can be traced deep into the animal world.

CHAPTER II.

"Instinct."

FROM this standpoint the study of the intellectual capabilities of animals is of a far other and of a far deeper significance than of old, when, as already stated, it was generally regarded and treated rather as a pastime, or as hunting out anecdotes for amusement, or as adorning theological and teleological disquisitions, than as a scientific system. For if it be true that organic gradations form an unbroken series, and that man himself is compelled to trace his origin to a set of lowly organised forms, as is maintained by the theory of development and descent—now more and more accepted—then is it further clear that not only the bodily, but also the mental powers of man must have the same origin, and intellectual development must be regarded as a common quality of organised matter. Comparative anatomy, *i.e.*, the study of bodies, which we have long followed, must necessarily have beside it comparative psychology, the study of minds; indeed the former must find in the latter its true completion. "The study of the special mental science of man, which is the groundwork of universal mental science," says a more recent writer on the subject (T. Bignoli: " On the fundamental law of intelligence in the animal kingdom," p. 25), "requires this foundation. . . . . a comparative psychology of animals is unattainable . . . . if this same psychical power is not studied in the whole of intelligent life. The animal kingdom is, so to speak, without a head, and man without feet on which he can stand. . . . . The science of comparative anatomy and psychology is therefore senseless and cannot be understood, unless it be crowned with the still more fruitful science of the study of mind in animals."

This claim and this deduction are so clear and irrefragable

that Darwin himself, in his famous book on the "Descent of Man," did not begin, as might be expected, with the comparison of anatomical or physiological observations, but with a digression upon the gradual development of intellectual powers in animals; he well understood that if he were to succeed in proving the animal descent of man, or even in showing it to be possible, the objection would still remain that man, so far as his mental faculties were concerned, was nevertheless entirely and essentially different from the rest of nature. And, although the great naturalist had at his command comparatively small and poor materials (in reality they are far more numerous and more convincing), it was yet not difficult for him to find and point out the rudiments, the roots, in animals of nearly every intellectual and moral capacity found in man. For the rest, this digression, like everything written by Darwin, is admirable, rich in facts and in striking deductions and observations, although Darwin here, as in his other writings, uses always the mischievous term "instinct;" this word leaves room for so much misapprehension, that it should be entirely avoided in scientific books, for, as Dr. Weinland well points out, it is nothing but a lazy way of escaping from the laborious study of animal intelligence.

"It is as though there were some witchcraft in the word 'instinct,'" says Umbreit, in his "Science of Psychology" (1831); "for with the phrase 'It is instinct,' all search after the manifestations of the intelligent life [of animals] is stopped as if by an anathema."

"The distinction between intelligence and instinct," says J. Franklin, "is now given up by all schools which have examined facts. There is intelligence in animals and instinct in man."

"Instinct," says Dr. F. C. Noll, in an admirable treatise on the manifestation of so-called instinct (Frankfort, 1876), "is merely an empty word, a veil for our ignorance or indolence."

Darwin does not, of course, use the word instinct in the old sense of an inexplicable and unchangeable impulse, springing from an unknown source; he uses it to describe an inherited influence or tendency, originally acquired through adaptation or natural and sexual selection, and habits or capabilities of thought transmitted from generation to generation. This is the only sense in which the word can be now used by educated people; in this sense, clear manifestations of instinct play a very important *rôle* in man's life, as in that of animals, although its influence is stronger and more apparent in the latter. It has, therefore, always aroused man's wonder, and led him, since he could not explain its origin, to the useless and foolish notion of an " instinct; " just in the same way, our ancestors, as they could not explain the rising of water against gravity into an exhausted receiver, ascribed it to nature's *horror vacui*, abhorrence of a vacuum, or as uneducated and thoughtless people now refer the origin of life to a special " vital force." That nothing is made clearer by such subterfuges, but that the whole question is made darker by the encouragement thus given to obscurity and mental sloth, must be seen by all. Shakspere jeers very bitterly at instinct when he makes Falstaff say in excuse for his senseless cowardice: " Instinct is a great matter; I was a coward on instinct."

Man, indeed, by a careful study of animal intelligence, by experience and observation, is led step by step to note manifestations and facts, and thus the conception of instinct is shown to be inconsistent, and is shattered in pieces ; that is, if instinct be used in the generally accepted sense of an inborn, inherited, unchangeable and therefore never erring impulse of nature, bestowed for the preservation and propagation of animals, the manifestations of which are unconscious and purposeless. Accurate study shows that the greater part of the actions attributed to instinct may be understood in a very different and more natural way, as arising sometimes from true consideration and free choice, sometimes from experience, instruction and training, from practice or imitation, sometimes from a special development of a sense, as smell, sometimes from habit and organisation, from reflection, etc., etc. For instance, when the caterpillar uses the thread given him by nature for spinning his cocoon, to let himself down from a tree and to escape thereby from his pursuer; when caterpillars, shut up in cases, tear off and use for their chrysalis the paper with which these cases are lined ; when the toad eats a great quantity of ants which it cannot digest, because they suit its taste, although it knows that it is thereby drawing upon itself pain and sickness; when bees greedily eat honey mixed with brandy, although it makes them drunk and at last unable to work; when birds which build near human habitations habitually utilise remnants of human work, strings or woollen threads, in building their nests;* when, as has been observed by G. H. Schneider, lobsters in captivity use bits of linen and paper to hide under, instead of pieces of plants, although they use only the plants if they are given a choice between the two; when a bee, if a ready-made set of cells be given it, leaves off building and carries its honey into the artificial cells; when a bird escapes the labor of nest building by using a prepared nest-box or by usurping the nest of another bird; when, after the same fashion, ants rob others of their nests and establish themselves therein instead of making their own; when swarms of bees, instead of gathering honey, remove the stores of another hive; when many animals imitate the voices and cries of others which chance to be in their neighborhood, with the object of defence or of enticement; in these, and in a thousand similar cases, the relation of which would fill a whole volume, instinct is an impossible cause or motive of such actions. Why do animals of the chase more than other animals fear men who carry guns? Why does a hound tremble if you point a gun at him? Why do large animals fear man more than do small? Why do old birds build better nests than young ones? Not from instinct, but from experience. Why does the fox rob the henroost at a time when he knows the master or the servants of the house are away or at table? Not from instinct, but from reflection. Why does the dog hide the remains of his food that he may use it later? Not from instinct, but from forethought. When the chamois (like so many other animals) set sentries to warn them of approaching danger, they cannot have learned to do so by instinct, for chamois

---

* The Oriole, *Oriolus galbula*, according to Ponchet, now never builds a nest without using scraps falling from man's industries, such as threads, yarn, strips of leather, or even cords, watch-chains, etc. In a wood near Mayence, where paper chases often take place, it builds its nest (according to W. von Reichenau, "On Birds," 1876, p. 62) by weaving the scattered bits of paper, fragments of letters, and journals, with the long grasses of which its nest used to be composed. The same thing has been observed in the so-called "Baltimore Bird,' *Cassicus Baltimore*, in North America

hunters, against whom alone such farsighted prudence can be directed, are much more modern than chamois!

As already pointed out, unchangeability is one of the great characteristics of instinct, and further, it cannot and must not blunder in guiding animals for their good. But there are countless instances in which instinct not only blunders, but is excessively changeable and subject to alteration under changing circumstances and life conditions. The fleshfly, whose maggots live on putrid flesh, very frequently lays her eggs on the leaves of the *Stapelia hirsuta*, a plant which is sometimes cultivated in hothouses, and which has an odor of putrifying flesh. And flies, in like fashion, take rotten leaves for carrion, on account of their smell, and lay their eggs therein, although in both these cases their progeny must perish from want of food. Here the animal is not led by instinct, but is misled by smell, which, looking at the circumstance that it has been born and bred among such smells, is not surprising. Instinct blunders in the same way, if, as Brehm relates ("Animal Life," 2 ed., IX., p. 16), the pine-moth, whose caterpillars eat pine leaves, lays her eggs on oak trees growing in the neighborhood of pines; or when, as Brehm, the father, ("Life of Birds," p. 247) has seen, a siskin leaves its half-built nest, because it has found, during its work, that the boughs of the tree on which it has built grow too closely, and do not give it room enough; or when the swallow takes wet street-mud for clay and builds therefrom a nest of brief duration; or when great waterbeetles fling themselves with suicidal force against the glass of a hotbed, thinking it water; or when birds try to drink shining potsherds; or when young birds, who are old enough to begin to eat alone, chirp at the food in the hope that it will then of itself come into their mouths; or when grazing animals eat poisonous plants with which they are unacquainted, etc., etc. "In the valley of the Ahn," says the distinguished clerical observer, Snell ("Zoological Gardens," vol. iv., p. 61), "from Michelbach to Langenschwalbach, and in some neighboring valleys, grows very plentifully the stinking Hellebore, *Helleborus fœtidus*. The sheep of the neighborhood know very well the poisonous properties of this plant, and never touch it, although they graze constantly on the hills and cliffs where it grows. But as soon as strange sheep, from a place where this poisonous plant does not grow, come to Hohenstein, they eat it without suspicion and poison themselves. Very many sheep bought in other places have fallen victims in this way. There is then no instinct warning the sheep; they eat the flowers and flower-buds of the hellebore, which are absolutely fatal to them, more greedily than the leaves, which, as a rule, only make them ill. This is the more noteworthy because the sheep has not degenerated by being stall-fed, but lives in a half-wild state."

According to the same observer (" Zoological Gardens," vol. iv., p. 60, etc.), dogs learn to know their food by experience, partly by instruction from the parents, partly by their own attempts, or by tasting. When he threw to his pigeons oats mixed with the, to them, unknown buckwheat, they eat only the first and left the other. But when they had gradually and forcibly been habituated to the latter, they learned to like it so much that they left the oats and picked out the buckwheat. Newly caught birds must at first have food they know mixed with the so-called cage-food, or they will die of hunger in the midst of the greatest abundance. Nay, each individual animal makes his own experiments and his own discoveries in respect to his food, and of this our observer brings a number of interesting examples, and proves that certain single individuals are able by example and imitation to rapidly spread habits acquired in this way.

That instinct, and even the strongest instinct, that of food, can change in the most radical manner, is shown by the example given by Dr. F. C. Noll; a parrot, *Nestor notabilis*, a native of the New Zealand Alps, which had lived on flowers, and berries, and chiefly on insects, amused himself with the meat barrels of the colonists, and then, once accustomed to flesh food, took such a fancy to it that he would not spare the sheep-skins hung up to dry, and finally picked such big pieces out of the flanks of the living sheep that the poor creatures died from exhaustion. Among other things, as Brehm relates (" Animal Life," 2 ed., vol. iv., p. 169), a flock of these birds pecked a sheep actually to death by their united exertions, in order that they might dine off it more comfortably. A similar fact is told by Snell of a black cockatoo in Java, which learned to kill and eat porpoises; he tells us also of the hen of a pair of ravens, which acquired the magpies' habit of plundering birds' nests, while the cock bird did not thus misbehave himself; of a coal titmouse, kept by him in captivity, which in default of his usual worms, pieces of meat and so on, killed a robin, eat it bit by bit, and thenceforward continued to murder and feast on smaller birds; and lastly of a Lammergeyer, *Gypaëtus barbatus*, which in some countries and places is a bold thief and bird of prey, and in others, compelled by circumstances, is a regular feeder on carrion. Brehm tells us also of a parrot which killed and eat birds of its own species or smaller birds. Reclam ("Mind and Body," 1859, p. 300) watched flesh-eating squirrels and rabbits, which latter gnawed like dogs bones which were thrown to them, although they were never short of vegetable food. Yet more striking is the instance recorded by Darwin of those island cattle, which on islands where pasture failed them, accustomed themselves to a fish diet—an observation which exactly coincides with those of Brehm on cattle in the extreme north of Norway, which were fed on cooked fishes' heads, and where the frames on which the haddocks were drying had to be guarded against the cows, because these were wont to feast without ceremony on the half-dried fish. Even horses will, under some circumstances, become flesh-eaters, as in a case observed by Dr. R. A. Philippi, at St. Jago, in Chili, wherein two riding horses belonging to a Mr. Nicholas Paulsen would snap up and eat young chickens and pigeons, pulling the latter out of their nests in a low clay wall.

Another instance of a changed instinct: The so-called hole-beaver, which, in consequence of the snares of the hunters, cannot live in societies any longer—as for instance in France and Germany—has changed from a social to a solitary animal; instead of its famous structures, the dams built in the rivers (as in America), it digs itself simple holes in banks, and barricades them, at most, with some logs of wood. Thus, merely from stress of circumstances, and in opposition to its supposed instinct, it has become a digging and hole-dwelling animal, instead of a building one living in the open. In a similar way the sable, as Radde relates (in Brehm, vol. ii., p. 60), is, in the Baikal mountains, owing to the hunters' traps, an animal that hides in holes and fissures, while in the Bureja mountains it seeks hollow trees and avoids rocky crevices. So also with rabbits, which for several generations have lived where they could not burrow; the descendants of these non-digging rabbits have lost the love or desire, formerly so strong in them, of digging holes (Reclam). Darwin notes that the Egyptian pigeons have changed, owing to the constant neighborhood of the river, and have learnt to sit on the surface of the Nile and drink from it like water birds. The wolf, as Noll points out, is, in places where he is supreme, a bold impudent robber, whereas he becomes a timid beggar when he approaches cultivated places. The blackbird, *Turdus merula*, was formerly a very shy bird, but in gardens and in the suburbs of towns wherein it is spared and welcomed, it has become so confident that it builds its nest with the help of bits of paper amid foliage where it is daily visited by man, and takes its food from the windowsill in company with sparrows. The ringdove, *Columba palumbus*, another formerly shy bird, in Pfannenschmied, in East Friesland (" Monthly Journal of the Saxon-Thuringian Society for the Study and Protection of Birds," March, 1876), where a suitable breeding place was wanting, nested in the direct neighborhood of human abodes, and even in the most frequented streets of the town. Dr. G. Jäger ("In the matter of Darwin against Wigand," p. 240) relates that he observed a six-weeks old kitten, which possessed to so high a degree the utterly uncatlike love of the bath, that he found it one morning in his pan, and that no water vessel was safe from it. The same observer noticed that each year a number of young ones appeared among his oak-caterpillars, which entirely lacked one of the most powerful instincts, that of eating; they would wander about restlessly over the food, until they finally starved to death. In like manner we often find suckling animals which lack the instinct of sucking, and calves, which are generally taken away after birth from the milch-cows, gradually forget how to suck.

The silver sea-gull, *Larus argentatus*, according to Audubon, sometimes, against its instinct, nests on trees, and certainly the older birds do this on White Head Island and the neighboring islands; they formerly nested in the swamps, and they must have changed after they found out that their eggs were annually taken from them by the fishermen:

some of the younger birds, however, still nest in the swamps. Bees, taken to the Barbadoes, lose their instinct of storing honey, because they find food enough all the year long in the sugar canes, while those in Jamaica, where the rainy season prevents them from flying out for several weeks, keep their instinct (Perty, "On the Intellectual Life of Animals," 1876, p. 41).

People are wont to say of the cuckoo that its instinct leads it to lay its eggs in strange nests. How comes it then that the American cuckoo does not possess this instinct, but sits on its own eggs? or that there are also other birds which sometimes or occasionally lay their eggs in strange nests, in order to save themselves the trouble of sitting? Or what has instinct to do with it, when the ostrich, like many other birds, leaves the business of hatching its eggs to the sun, during the day, and only covers them with its body during the cool of the night? Or when the same bird, acting in this way in Senegal, never leaves its eggs day or night at the Cape of Good Hope, where the warmth of the air is less? Or when geese and ducks, in our moderate climate, leave their eggs for awhile without any care, while the same birds in the polar regions in such a case cover their nests with feathers as a protection against the cold? In a similar way the Canadian muskrats, translated into a warmer climate, cease to build their neat warm nests, and content themselves with a mere hole in the ground.

One of the very famous instincts is that of the honey bee, building six-sided cells; a propensity which, as will be shown later, has been evolved in the most natural way. But that this is not unchangeable, is shown by the fact that bees guide themselves by circumstances, and give the cells another or an imperfect shape when insuperable obstacles are in their way. Darwin has also often observed, that cells which had to be built in a corner or in some inconvenient place, were repeatedly pulled down and built up again until the architect was satisfied. The examples of the change and of the improvement by circumstances in the building instinct and other tendencies—a fact which is utterly opposed to the conception of an instinct—are so numerous in the insect world that Blanchard, in his great work on the changes and habits of insects (Paris 1868) wrote:

"Instinct by itself ought always to guide individuals of the same kind to the same work. But suppose there are difficulties in the way of the completion of the work? The individual removes the hindrance; he selects the best place for his dwelling; he guards against accident, he meets danger. Sometimes he gives way to sloth, and builds no house, but steals someone else's, and scarcely ever improves it. The insect, believed by man to act as a machine, gives proof at every moment that it takes note of the situation in which it finds itself, and of a crowd of accidental circumstances which it was impossible to foresee. But to take note of a difficult position, to improve it, to make a choice, to take a thing in order to save oneself trouble, to be lazy when one is made for industry, is this instinct? Impossible!"

In the same way speaks C. Menault, in his admirable book on "Intelligence in Animals" (Paris, 1872, p. 114): "What! creatures that have faculties, that feel, remember, and compare their feelings, that express themselves in a more or less distinct fashion, but ever in sympathy with their emotions of joy, grief, anger, or passion—such creatures have no intelligence? By God, I should then like to know what intelligence is!"

Note also what Wallace, the thought-colleague of Darwin, brings forward in his "Natural and Sexual Selection" (1870), on the many changes and alterations of a very powerful instinct, namely the nestbuilding instinct of birds, in consequence of change of circumstances. "The orchard oriole of the United States," says Wallace, "offers us an excellent example of a bird which modifies its nest according to circumstances. When built among firm and stiff branches the nest is very shallow, but if, as is often the case, it is suspended from the slender twigs of the weeping willow, it is made much deeper, so that when swayed about violently by the wind the young may not tumble out. It has been observed also, that the nests built in the warm southern states are much slighter and more porous in texture than those in the colder regions of the north. Our own house-sparrow equally well adapts himself to circumstances; when he builds in trees, as he, no doubt, always did originally, he constructs a well-made domed nest, perfectly fitted to protect his young ones; but when he can find a convenient hole in a building or among thatch, or in any well-sheltered place, he takes much less trouble, and forms a very loosely built nest." Our little golden-crested wren also changes his nest to suit circumstances: he builds a simple cup-shaped nest, where a natural canopy of sheltering leaves is to be had, and in more exposed situations he builds a spherical nest, with an opening at the side.

The little bittern, *Ardetta minuta*, according to Brehm, ("Pictures from the Animal World," p. 158) is wont to build his nest on old reed straw, or on reeds standing over the water, or on an over-hanging willow. If it floats, as at other times, on the water itself, it is then only loosely attached to the reed, so that it may rise and fall with the water.

A remarkable instance of a change of instinct is recorded by Wallace as lately happening in Jamaica. "Previous to 1854 the palm-swift, *Tachornis phœnicobea*, inhabited exclusively the palm trees in a few districts in the island. A colony then established themselves in two cocoa nut palms in Spanish Town, and remained there till 1857, when one tree was blown down, and the other stripped of its foliage. Instead of now seeking out other palm trees, the swifts drove out the swallows which built in the Piazza of the House of Assembly, and took possession of it, building their nests on the tops of the end walls, and at the angles formed by the beams and joists, a place which they continue to occupy in considerable numbers. It is remarked that here they form their nests with much less elaboration than when built in the palms, probably from being less exposed."

Birds especially, which build their nests on or in human dwellings, buildings, etc., have largely changed their method of building to suit the changed circumstances, and have so changed it as to save their work and pains. In the same way the lake dwellers, who formerly lived 'over the water, carried on to dry ground their fashion of building on piles, although the fashion had no longer any reason; while, on the contrary, over the whole American continent the houses of the natives are built in a way which was suitable to the cold climate whence the Indians originally came. Wallace finds himself led by his investigations to say: "I believe, in short, that birds do *not* build their nests by instinct; that man does *not* construct his dwelling by reason; that birds do change and improve when affected by the same causes that make men do so; and that mankind neither alter nor improve when they exist under conditions similar to those which are almost universal among birds."

Quite irreconcilable with the instinct theory is the fact alluded to, that so many birds instead of building their nests, according to instinct, take possession of ready-made nests or holes, and make themselves at home therein. Thus the puffin, *Fratercula arctica*, uses rabbit holes, after it has turned out the rightful owners, and only decides to erect an underground building on its own account if the other does not suit it. The swift, *Cypselus apus*, on its return in spring time from warmer climes, ejects from its wonted resorts sparrows, starlings and redstarts, and settles itself down at home, using for its own purposes the stuff gathered together for nests. The eggs and the unfledged young of the legitimate owners are thrown out without mercy. That the nest-building of birds does not rest on mere instinct is further shown by the fact that, as already mentioned, old birds build better nests than do young ones, and that young birds brought up alone in cages either build very bad ones or no nests at all.

CHAPTER III

"INSTINCT"—*(Continued).*

IT results from all the facts stated, that as regards the instincts of birds a mass of tales and fancies have been generally and hastily believed, which have been carelessly repeated from one to the other without any trouble being taken to verify them, which really have not been proved, and which are shown by accurate observation to be either quite erroneous or at least very much exaggerated. As a common-place example of this kind may be cited the well-known instances, generally esteemed as proofs of instinct, as to poultry and ducks. It is said that young chickens, when they have developed to their full size in the egg, break their shells, quit them, and at once stand on their feet, run, and peck grain and insects from the ground; thus arises a whole series of very complicated motions, directed to a certain aim, without any teaching, example or experience coming in. Just the same is told of young ducks, which in addition give an especial proof of instinct, in that as soon as they leave the eggshell, they run to the water and swim about therein. This last feat is said to be done by young ducks hatched under hens, which therefore cannot be led to swim by any maternal allurement, and the poor foster-mothers are said to stand in despair, because they see their nurslings torn from their protection, and are not able to follow them.

This all seems so natural that it is generally accepted without demur, and would indeed, were it true, scarcely seem to leave doubtful the existence of instinct in the earlier accepted sense. But, in reality, matters are very different. That which occurs immediately before the escape of the chicken from the egg does not depend on the independent action of the young bird, but takes place in quite mechanical fashion, and as the result of a series of unconscious, or so-called reflex movements, which are caused by the fact that from twenty-four to thirty-six hours before hatching the chicken begins to breathe, and at last requires more air than can pass through the shell. Hence arises a great danger of suffocation, and in consequence of this strong reflex action takes place, by means of which the chick is caused to strike, or push violently against the inner wall of the shell with a sharp bony point formed on the beak, and the whole body is stretched and extended. Pressure is caused by the natural growth of the body within, and the breaking of the shell cannot long be delayed.

But when the chicken is out of its shell it is far from running about and picking up corn. It generally lies for about two hours helpless on its stomach, and neither eats nor pecks, even though a grain of corn be placed in its beak. It then begins to make feeble attempts to move, in which it at first uses its wings just as though they were crutches. It gets up, falls down again, falls down and gets up, so that its motion looks more like slipping than running. If a noise is made near it, for instance if anyone knocks a table with the finger, it turns to the side of the noise, and this is not surprising, as its ear has already to a small extent been used within the shell. During the next six hours, the chick gradually gains strength and practice enough to run, and it also begins to peck at the ground, but blindly and senselessly; for it pecks at everything which reaches its eye, such as little lumps of earth, heads of nails knocked into the boards, grains of sand, and glass-beads, and even at mere bright specks which may have been brought on to the table or slate shelf among the chalk. This is done also by grown poultry, which are frequently seen to peck at the ground although there is nothing there to pick up. They also, like the chickens, peck at pieces of chalk, until experience has taught them that it is useless. Even poultry from which the cerebrum has been removed, and which are therefore without consciousness and without feeling, strike mechanically with their beaks on the ground without picking up corn, just in the same way as human babies try to put into their mouths whatever is given them. It ought not, therefore, to be surprising that chickens should do the same, especially when imitation of the pecking mother comes into play. That imitativeness and the teaching of the mother have a very large share in the whole concern is proved by the fact that the whole recorded process, until the chick is able to run and feed itself, takes only from five to eight hours, if it remains with and under the care of the mother, while it takes from eight to sixteen hours if the chick be taken away from the mother immediately after hatching.

The ducklings begin life just as do the chickens. They easily fall over on their backs and cannot get up without help. At first they neither eat nor peck, even if their beaks are pushed into moistened meal. As to their immediate running to the water, this is so little true that they far more often anxiously try to get out of the water when they are thrown into it. They do not drink of their own accord, but learn to do so gradually if their beaks are held in the water. When they are given drink in a cup they take it very clumsily, often touching the edge of the cup with their beak instead of the water. When they have learned to drink, they peck at a shining slate-shelf as if it were water. That other birds also have to learn to drink is shown by an observation kindly sent to the author by Frau Ruge of Schwerin. Frau Ruge saw a hen-pigeon with three newly fledged young take them to the edge of a water-tub and teach them to drink with great trouble. The whole process, the incidents of which are very well told by Frau Ruge, lasted a full hour.

Ducklings, like chickens, turn to the side where a noise is made; for example, to the side where anyone is speaking or where other ducklings are chirping. They gradually learn to walk by constant stumbles and falls, and also peck at nails, bits of chalk, etc. They behave just in the same way in the open. If they are brought to a pond they will go and drink, but will not go in. If they are placed in deep water they try to get out of it as quickly as possible, and therefore make active movements with their legs, which necessarily propel them, and, as an animal cannot sink, look like the movements of swimming. Dr. Stiebeling of New York, from whose admirable work " On the Instinct of Poultry and Ducks" (New York, 1872,) we have borrowed the above observations, noticed this in birds of one or two days old, and even in those of eight days of age when they had not left the coop, and, indeed, especially in these. Only gradually do the little creatures accustom themselves to losing the solid ground from under their feet. If the ducklings be hatched under a hen they take much longer before they become used to the water than if hatched under a duck, for the latter, like all water-birds, takes her young on her back, and, swimming out, she shakes them off into the water. When the ducklings reach dry land again, they shake themselves and try to clean the water off. This, as well as the fact that if they are given milk instead of water exactly the same incidents are observed, proves that we ought not to speak of ducks having an instinctive love of the water, even though there can be nothing surprising in their having an inclination to an element in which their parents and ancestors have been at home from time immemorial.

This thoroughly coincides with the facts kindly communicated to the author by Herr Julius Tapé of Szegzard, (Hungary) who lived long by the Danube, and had over and over again seen that the goslings feared the water, until they had learned to swim, and were, it might be said, cheated into it by the parents. When the young are old enough to be taken into the water, the old ones bring them to the bank. The gander goes in front, cackling continually, while the hen pushes from behind, cackling in the same way. After a very short trial-swim the goslings are brought back to the bank, and these trials are repeated from day to day at increasing length, until the young go into the water of their own accord. This gentleman has also observed that when geese wanted to swim over from the opposite bank, they never started from a point exactly opposite, but took to the water much higher up the stream; and they knew so well how to calculate that when the wind did not interfere with them they landed exactly opposite their owner's house. If by chance a steamer was in the neighborhood, they quietly waited until it had passed.

We cannot, then, talk about an innate love of water with reference to these animals either. This may be said rather of young turtles, of which it is told, and apparently truly, that as soon as they escape from the eggs hatched in the sand of the sea-shore, they run to the sea, and repeat this even if they are forcibly turned back, or driven in an opposite direction. But here it is clearly nothing more than the scent of the sea which causes this action, and it is well known that the sense of smell is far more highly developed in most animals than it is in man. It is certainly not very marvellous that young animals should follow the scent of an element in which their parents and ancestors have lived for unknown ages. This is yet more plainly seen in the far-famed instinct of metamorphosed insects, which always lay their eggs on places suitable for the nourishment of the escaping caterpillars, without recognising these places absolutely by their own sight. Here, without doubt, they are guided in their action by the sense of smell, so highly developed in insects, and perhaps also by some kind of memory from their caterpillar or maggot condition. "The *Sphinx Euphorbiæ*," says Noll, (p. 15,) "knows the milkwort by its shape more than by its smell. And why should it not? Is it not the particular flower with which it is thoroughly acquainted? Has it not itself in its growth, in its caterpillar stage, lived always on this plant, imitated its color, and nourished itself on this alone? Has it not built up its body out of its tissue, taken into itself its volatile oils and its alkaloids? Do we not note how the land-fly, which kills the caterpillar of the swallow-tail butterfly, *Papilio machaon*, by stinging it behind the head, spreads a strong smell over the carrots on which it lives? It is also known that the blood of many insects, and especially of larvæ, smells of the plants on which they feed. The dancing butterfly, seeking honey as its food, certainly knows what it enjoyed in its youth and where it lived; for even though its form be changed, though its intestines with their peripheral nerves have disappeared with its transformation, yet the chief part of its central nervous system remains, as has been proved in the transformation of caged insects, and may therefore, have well preserved the youthful impressions, which in man are indeed the most lasting." So the moth finds its way into locked-up cupboards of clothes, which it has never seen, merely by the aid of smell, and when we, as a protection against it, put strong-smelling substances, such as camphor, turpentine, etc., among the clothes, we do so to conceal the smell of the woolstuffs by a stronger smell, and so to deceive the scent of the moth. Of what great and marvellous performances the smell of insects is capable is proved by the well-known fact, that if, in the midst of an inhabited place, we expose a female moth in the window, the male will in the shortest time be led to it by scent from places miles away. If such a female is even kept shut up in a room, the male, led by the smell, will come to her *viâ* the chimney, as Mr. Davis observed in England in the case of the *Sphinx populi*. The same observer, later, on the occasion of the emergence of the females—of the *Phalæna bucephala* and *P. salicis*—found that the window of his study was besieged by males who had been attracted by the smell. Herr Lüders, of Attenburgh, relates that on accidently opening a desk, in which a female Dispar had emerged *a year before*, a male Dispar flew in through an open window, and did this several times without noticing anything else in the room (Reclam, p. 312 and 313). Darwin ("Descent of Man, ed. 1875, p. 252) records after Blanchard (among a number of similar observations), that M. Verreaux, in Australia, carried the female of a little Bombyx in a box in his pocket and was accompanied by such a crowd of males that over 200 followed him into the house.

In this way, when they are closely looked into, are explained in the most natural fashion many cases of supposed instinct; in others, as in those of poultry and ducks, the current stories are inaccurate. Doubtless such an exact investigation as that of Dr. Stiebeling on the last-named instincts, would show a similar result in apparently very convincing cases, if people, instead of accepting stories on trust, would take the trouble themselves to test and observe. "No one," says Wallace, "has ever taken the eggs of a bird which builds a complicated nest, hatched the eggs by steam or under some alien mother, and then brought the birds into a large aviary, or covered place, wherein they should have materials and opportunity to build a nest like their parents, and then observed what kind of nest they would build. If under these strict conditions they chose the same materials and the same situation, and constructed their nest in the same way and as perfectly as their parents, that would show instinct. But now this is only assumed, and as I shall further show, is assumed without satisfactory grounds. So, too, no one has ever taken the larvæ of the bee out of the hive, kept them apart from other bees, brought them into a great enclosure with many flowers and abundant nourishment, and then observed what kind of cell they would build. But until this has been done no one can say that bees build without instruction, no one can say that with each new swarm some bees do not go which are older than those of the same year, and which may act as teachers in the building of a new comb," etc. That bees and ants actually teach each other, and that the younger do other and easier work than the older, and have quite other instincts, will not long remain unknown to the reader of this book. The beavers also, about whose remarkable architectural skill so much truth and so much falsehood have been written, instruct one another; and if we read in nearly all books on Animal Psychology the well-known stories of the young beavers, repeated and used for the hundredth time, how they are taken from their mothers immediately after birth and brought up artificially, and with materials provided in their cage manufacture a dam after the most approved rules of beaver-dam-construction, then, without making the experiment, we may say decisively that the story is false, or at the best much exaggerated. The young beaver, impelled by his inherited tendency to build, may very well make attempts to construct a dam, as, according to P. Flourens (pp. 53 and 185), one of the animals brought up by Cuvier actually did; but that a dam was constructed by it according to all the rules of beaver architecture, without aid or instruction from its older companions, can be believed all the less, since the beaver-trappers expressly assure us that, according to their observations, young beavers stay no less than three years with their parents in order to be instructed by them. "The young ones," says Schmarda ("Intelligence in Animals," p. 207), " live until the third year with their parents, and are then sent to build for themselves." On the other hand, the animal has so much intelligence not ruled by instinct that it can turn its architectural ability to special account according to circumstances. E. Menault (p. 195) relates that a beaver, kept encaged at the Botanical Gardens in Paris, one cold night, when the snow came through the palings of his enclosure, built up his cabbages, fruits, and twigs of trees into an excellent shelter against the annoyance, weaving the tree twigs into the grating, and the snow itself serving for mortar.

" On all sides," says Espinas, the distinguished author of the " Treatise on Animal Societies," " it is admitted, even by those least favorably disposed to animals, that the latter act individually, and from motives changing according to circumstances. This is one of the greatest results of comparative psychology in the second half of this century, and the time is near when no one will any longer dispute it."

There are certainly many actions of an instinctive kind, but—if they cannot be explained as the result of reflection, imitation, habit, instruction, experience, and consideration, or from a specially fine development of a sense, and other peculiarities of organisation—they depend in each case, as has been shown, on propensities inherited from parents, or mental habits and capacities, or, to put it anatomically and physiologically, on inherited predispositions of the brain and nervous system for certain psychical function-grooves— briefly, on what may be called inherited memories. These tendencies and habits, perhaps even ideas of a certain sort, are acquired by parents and ancestors during their lives in a definite manner, generally very gradually and in the course of very long spaces of time; these are transmitted, when they bring or have brought advantage in the struggle for existence to their possessors, and are handed down intensified from generation to generation. Artificial breeding may work in the same fashion as do the struggle for existence and natural selection in Nature. For instance, the well-known and oft-cited instinct of the hound or the pointer is nothing more than the lengthening by art and training of a brief pause, which all beasts of prey are wont to make at the sight or scent of their quarry just before springing at it, partly to gather up their forces, partly to fix attention as strongly as possible on the object to be seized. None the less must the pointer puppy, which has brought with it into the world a tendency to this action, inherited from its parents, be trained into a perfect and useful hound by education, punishment, and experience. In like manner is it with the instinct or inclination of sheep-dogs to herd sheep, or of the greyhound to course hares, or of the Newfoundland to rescue men, as also with the love and attachment shown to men especially by the dog, originally sprung from wild animals, apparently from wolves and jackals. The famous migrations of birds have only arisen gradually through a slow pressure of cold from the poles to the equator, and are now continued by transmission from generation to generation. Thence comes it that migratory birds kept in cages grow uneasy when the season of migration approaches, and beat their heads against the bars of their cages, although it may reasonably be doubted whether quite young birds, which had not yet taken part in a migration, and had been kept from all communication with their fellows, would show a similar desire. Should they even do so—as F. C. Noll (p. 22) maintains of the nightingale—without having had any possibility of communication with the outer world, it would then be only the inherited tendency to migrate which seizes the animal at the season at which his species migrates, as the result of an inborn and inherited organisation of the brain and nervous system.

It should the less surprise us that such inherited tendencies or habits are found in animals since they are not absent in man; as many of the phænomena of intellectual life in men show such a great likeness to those of animals, it becomes necessary, if people will insist on the existence of instinct, that the same explanation shall be given of the former as of the latter. But indeed there is no such thing as instinct, in the earlier application of the word, but there is an unbroken series of intellectual gradations from the lowliest animal to the highest man. "Instinct," says Lindsay ("The physiology and pathology of mind in the lower animals," 1871), "is not something separated from and opposed to reason, but is rather a necessary concomitant of the latter. Instinct and reason are merely different stages of development, or differing manifestations of the same capabilities, or of the same class of phænomena. They run into each other by such imperceptible degrees, that it is impossible either to draw a decided line of demarcation, or to find a clearly distinctive character. Instinct, like understanding and reason, is found both in men and in animals, although in varying degrees or manifestations. It is very difficult to distinguish innate from acquired capabilities, or to separate the results of intuition from those of experience. That which is an acquired capacity or property in the parents, becomes generally an instinct in the following generations, when custom has set on it its seal."

And the same author ("Insanity in the lower animals") says again: "I do not doubt that much of that which is called instinct in animals is exactly the same as that which in man is called reason, and has just as good a right to this title; while on the other hand, that which is called reason in man is the exact counterpart of that which is called instinct in animals, and ought so to be described. Little is known of human instincts; and indeed it is an expression which, whether it be said of animals or of man, has hitherto served rather as an *asylum ignorantiæ*, or a harbor of refuge for ignorance, and a real hindrance to investigation."

"Do not let us speak of blind instinct," says Michelet ("The Life of Birds," p. 3); "we shall see how this clearsighted instinct modifies itself to suit circumstances; in other words, how these beginnings of reason are very little different in nature from reason in man."

E. H. Morgan, in his well-known book on the "American Beaver and his works," (p. 275) puts the matter yet more forcibly: "The expression 'instinct' should be quite surrendered as an explanation of the intelligent action of animals. It is a metaphysical invention to make an essential difference between human and animal understanding, and it is quite impossible to explain thus the phænomena of animal intelligence. . . . . Animals possess an intellectual principle which does them the same service as that of men to men. . . . . And as we are not able to make a difference in kind between the phænomena of emotion, will, thought and reason in men and in animals, we therefore come to the conclusion that the difference is one of degree and not of kind. . . . . Healthy human thought—which represents far more than is generally admitted, the highest kind of human understanding—has never cordially accepted the speculations of metaphysicians as to the intellectual capabilities of animals. On the contrary, it has always been inclined to recognise in animals the existence of a rational thinking principle, analogous to that of man."

Many strong proofs of the existence of instinct in man may be brought forward, using the word in the narrow sense defined above. Only take, for instance, the instinctive lust of murdering and destroying his own kind, developed in individuals through the endless and bloody wars in the early ages of humanity, and by the struggle for existence, and continued indeed almost in its original strength in rough or half civilised people, while among civilised nations it has been brought within definite limits by law, custom and reason. But that in spite of these it exists in its old power in the breasts of individuals—and of very many—and is only prevented from bursting out by these restraining forces has, unfortunately, been sufficiently shown by terrible outbreaks of violence, both private and public, under special circumstances. Long centuries of peace and of usages, or political and social arrangements, directed to the common peaceful happiness and prosperity of all, will alone be able gradually to eradicate this terrible propensity or instinct of man's nature, largely developed by the inheritance of thousands of years, and thereby create a better and a happier world than the present. But a longer pursuit of this theme and of the weighty thoughts surrounding it would lead us too far beyond the scope of this work; before passing to this we will, therefore, only say a few words on the importance of information on animal intelligence to those Societies for the Protection of Animals, now so fortunately established and flourishing. These societies are one of the fairest signs of the humanity that rules to-day, although, on the other hand, it seems sad that they should still be needful, when six hundred years before Christ, the thoughtful creed of Buddha taught the same principles, and preached gentleness and kindness to animals no less than to man; nay, it even made it one of the duties of the faithful to build hospitals, not only for sick men but for sick animals. Christendom and Christian philosophy started from an entirely different point, making a division, a tearing apart, of soul and body, of man and animals, and necessarily thence arrived at principles of harshness and cruelty towards the latter. But the very existence of these societies proves that the better conscience of man struggles against such views: it proves that man does not see in animals mere breathing and vivified machines, driven by instinctive propensities, but that he recognises in these his brethren; to put it shortly, men to-day are better than their religion.

The results of these societies would, however, be much greater if knowledge of animals and of their intellectual life were fuller and better. Unfortunately, this knowledge, both in respect to wild and tame animals, is still very poor and imperfect, partly because very few have the opportunity of themselves studying and observing animals, partly because the perverted views of philosophers on the subject have turned people's heads. Those who know animals really and not only by hearsay will, as a rule, hold a very different opinion; they will see that the animal is—as the author of "Passages from the Diary of a Travelling Naturalist" (1855), so well says—not only physically but also intellectually and morally "a man in fragments," and that all the intellectual capacities of man, including the highest, are foreshadowed in animals, and hide therein their earliest beginnings. This important truth cannot be better expressed than by F. M. Trügel in his admirable "*Causeries sur la psychologie des animaux*" (Leipsic, 1856) in the following fashion : " The more man observes himself, and the more he studies with a critical eye the smallest details of the new and ever noteworthy phænomena of animal life, the more will he be penetrated by the great truth that animals, like men, think, will and feel. If we pass from the study of men to that of animals, we are astonished to rediscover in animals all that which we had before discovered in the innermost recesses of man's heart and mind. At each step taken in this unwonted sphere we meet surprise after surprise. Sense and stupidity, craft and simplicity, good and bad taste, kind heartedness and malice, gentleness and harshness, impetuosity and indifference, prudence and thoughtlessness, gravity and frivolity, courage and cowardice, modesty and boasting, unconcern and anxiety, fidelity and faithlessness, inclination and aversion, love and hate, candor and knavery, pride and humility, gratitude and ingratitude, refinement and coarseness, trust and mistrust, reason and folly, compassion and severity, prodigality and avarice, temperance and gluttony, hope and despair, wilfulness and pliability, obedience and rebellion, grief and joy, anger and patience, sloth and industry—in short, the dispositions, the feelings, the good and bad qualities of men, rise one after another out of the great ocean of animal life; there is revealed to the astonished observer the perfect image of our whole social, industrial, artistic, scientific, and political life."

That there is no exaggeration in these last words, and that just those things which we regard as peculiarly human— the formation of a state and a society in its smallest details, the art of building, husbandry, war, slavery, language, etc.— are represented to an almost incredible extent in the animal world, is nowhere seen more plainly than in those minute and little observed animals, which without noticing we crush by the dozen as we walk, and of whose marvellous instincts Darwin has lately treated so impressively. Who among our readers has not read Darwin's renowned book on the " Origin of Species," and studied with astonishment his observations on the so-called slave-making instinct of ants? The striking facts were indeed known long before Darwin, and were accurately observed and made public by the Genevan, P. Huber, at the beginning of this century. But great as was the interest aroused by these publications, they yet did not meet with the full and free acceptance from the educated world deserved by their nature, owing to the fact that man was still at that time quite under the yoke of the oft-named prejudice touching the intellectual life of animals. Now it is different, and Darwin's observations on the slavery-instinct have specially aroused great interest, because an institution is there concerned which has played, and partially still plays, a large part in human culture and development. But before we can speak of this we must first deal with other matters concerning this wonderful insect and its political, social and domestic arrangements and customs.

ANTS AND ANT LIFE.

D

CHAPTER IV.

General Characteristics.

THE ants, in psychical or intellectual development, stand without doubt, highest among insects, although the better-known bees, with their highly-organised societies, are unjustly allowed to challenge their position. Such a challenge might rather come from the Termites, the white ants of tropical regions (wrongly classed with ants), of whose manners and customs enough is not yet accurately known to enable us to make an exact comparison of their mental capabilities with those of ants. Even the European ant, although studied with the utmost accuracy by a circle of distinguished observers, is not yet known with the exactness merited by this marvellous insect, with its remarkable political and social arrangements; doubtless future observers will bring to light thereon many noteworthy and surprising facts as yet unknown.

It will doubtless surprise many readers to hear that such highly-developed intellectual powers and capacities are to be found in so low a grade of animal life, and perhaps this very circumstance will be enough to fill them beforehand with a certain distrust as to the facts to be hereafter related. But it should not be forgotten that the great subkingdoms or divisions of the animal kingdom are not arranged beneath, but beside each other, and that therefore the highest outcome of a generally lower subkingdom may be, and is, bodily and mentally, very much raised above the lower or middle grades of a generally higher one. Thus the most perfect of the Mollusca is much higher than the most imperfect of the Annulosa; and the most perfect of Annulosa is far higher than the most imperfect of the Vertebrata, although the Vertebrata in its highest members reaches the utmost perfection as yet possible. In fact, we may say without hesitation, that an ant, which represents the highest type of the highest division or class of the Annulosa, namely the Insecta, is by its whole organisation raised far above the lower classes of the Vertebrata, such as Pisces and Amphibia, and by its mental capacity mounts as far as Aves, and as the higher or highest Mammalia. Leuret, the distinguished anatomist and naturalist, says of the ants, that they stand highest among the Invertebrata, and that even among the Vertebrata, excluding apes and elephants, there is none to place above them. Their history, he says, is that of men. They have a distinct language, they build dwellings with windows, saloons, ante-rooms, partition-walls, pillars, beams, etc. They give battle, carry on sieges, make prisoners and slaves. They keep milch-cows, and show the greatest care for their children. If we were no larger than ants or bees, and they were as large as we are, they would perhaps regard us as small, certainly very rational animals, but yet of less importance than themselves.

Leuret's colleague and countryman, Professor Chr. Lespès, of Marseilles, who has also thoroughly observed the habits of ants, declares that they " deserve our thorough sympathy," and recommends the study of them as " one of the most attractive imaginable."

Professor A. Fée (" Études Philosophiques sur l'Instinct et l'Intelligence des Animaux," 1853) places the Insecta (and therein the ants in the first place) in the highest grade as to instinct, and in the third as to reason, of the whole animal scale.

No one has a better right to decide on the general place of ants in nature and in the scale of existence, than Dr. Augustus Forel, who, by his excellent work on ants of Switzerland (1874) has trodden worthily in the footsteps of his famous predecessor and countryman Peter Huber, who first described the habits and noteworthy customs of the ants of his own country in the year 1810,* and from whom all later writers have borrowed more or less. The French Blanchard, in his great book on the relationships, habits, and instincts of insects (1868) speaks of Huber's work as a " revelation." " It will ever remain as a monument of patience and acuteness; his account of the facts bears

---

* " Pierre Huber; Recherches sur les Mœurs des Fourmis Indigènes." Paris et Genève, 1810.

everywhere the stamp of a truth set forth with the greatest attractiveness ; and the simplicity of the narration does not veil the enthusiasm of the author for the little world he studied so minutely."

Blanchard himself opens his chapter on ants with the characteristic announcement, that although the lack of trustworthy observers in former times gave free play to the imagination, yet all inventions are utterly outdone by the reality which we now know.

D. B. Graber ("Insects," Munich, 1879) calls ants the coryphees or Primates of the insect-world, on account of their highly-developed intelligence, their enormous numbers, and their motto, "*viribis unitis.*"

Forel, for his part, sets ants so high that he regards them as being among insects what man is among Mammalia. This would be most suitable as applied to the ants of the tropics, although Forel's observations were confined to native ants. "The rôle played by ants," says Forel, "in Nature's economy in Switzerland is a very modest one, as compared to that assigned to them in tropical lands. The strength given to these little creatures by their unity and intelligence is there shown in a surprising manner; and the tales of travellers on this matter mount often to tragedy. The Brazilians are wont to say that the ants are the real kings of Brazil, for they exert therein an almost unchallenged rule." " There can be no doubt," Forel goes on, " that the ants are the most intelligent of insects. Not Huber alone, but also Ebrard, Swammerdam, Lepeletier, and the other authors and writers who have taken the trouble to compare their habits and customs with those of bees, find themselves compelled to award them the palm. Their architecture is less artistic indeed, but it changes according to place and materials, suits itself to circumstances, and knows how to utilise everything, while that of bees always remains the same. Bees have no great trouble with their larvæ, and content themselves with carrying food for them into the cells. But the ants are obliged to feed their young from their own mouths, to bestow on them lengthened care, and to carry them from one place to another if there be a change of temperature, and they have to do all this during several weeks, while the bee-larvæ only remain in the larval condition for five days. Further, the bee finds its own way out of its cocoon, while the ants generally (apparently always, L. B.) need the aid of their companions. Lastly, the slave-making, the keeping of Aphides as domestic animals, and a number of other traits in their habits, are proofs of their mental superiority over bees, whose customs are simpler and more uniform. Especially do they surpass all other animals by their social instinct, or tendency to association, which shows a kind of collective or co-operative reason; this is indeed so great that we are involuntarily reminded of the little hostile communities of primitive men, and are compelled to think that the union of the individual intelligence of the higher Mammalia with the development of this associative tendency would have sufficed to produce man, with all his capacities; a thought which Darwin has followed out in detail in the third chapter of his book on man. No other animal shows such remarkable proofs of this associative propensity as do the ants. Swammerdam (1637—1680) compared the societies of ants to the early Christian communities. And it may be said that the ants have shown us a specimen of Socialism in practice, carried to its furthest consequences. Labor is quite free and unenforced; there are no chiefs nor overseers. Each ant is ready at any moment, without any kind of compulsion, to offer up his life for the community."

" The similarity of ant societies to those of men becomes especially striking if we observe the reciprocal relations between the various colonies. There are wars, truces, alliances, plunderings, robberies, surprises, tactics, stratagems of war. Nothing fails there of all that which man is accustomed to see. Especially remarkable are the alliances, and the executions of enemies; also the truces, which, after repeated battles are arranged between two hostile tribes or colonies."

Ants also show a great resemblance to men in the development of their character. To their great attachment and self-sacrifice for the commonwealth and for each member of it, are united generally a hasty temperament, prone to furious anger, and an unquenchable hatred against all foreign or hostile colonies. Therewith are blended industry, perseverance, and often cruelty. Also gluttony is one of their characteristics, as will later be more exactly shown: and their love for a good meal is so great, that it is thus possible to restrain their otherwise unconquerable desire to fight. Nothing is more interesting than to watch this struggle of two passions. If honey, of which ants are known to be inordinately fond and for which they will generally leave all other food, be placed on a battle field between two contending parties, as for instance red and turf ants, some of the warriors will be seen approaching and tasting it. They never stay by it long, but quickly return to the fight. Sometimes these same ants will turn back longingly twice or thrice.

In some races (for example, *Lasius*, *Tetramorium*) gluttony conquers the love of battle. We may also sometimes observe hate of an embittered foe struggling with friendship for old comrades, or fear and self-sacrifice for the community contending together in their little minds. This is true especially of particular individuals. One ant will let herself be killed, rather than let go the pupæ which she holds, while another will let them fall and run away like a coward.

The same thing is also true of particular species or families. For while some are cowardly and fearful, others show a fearlessness and a courage that make them a real terror to other animals. "Nothing is more interesting," says Forel, " than to pour out a sack full of turf-ants on to a newly-mown meadow, and watch the fashion in which they take possession of the new territory. All the crickets fly, while the ants take possession of their holes; the grasshoppers, ground-fleas, springtails, seek to save themselves on all sides; the spiders, chafers, staphylini, leave their prey, in order not to become prey in their turn. Clumsy creatures, or those who have lost their legs, or those which have just emerged, are killed by the ants and torn in pieces. I have seen a troup of turf-ants, which owing to the enlargement of their colony were lengthening one of their tunnels, come across the nest of a wasp, *Vespa germanica*, which had been built in the ground. They promptly blocked up the opening, and chased out the numerous inhabitants, not without losing many of their own number. When the cockchafers prepare to creep out of the earth in the spring, the turf-ants often crawl into the little holes which are not yet large enough to let the cockchafers pass, and kill them. Caterpillars, lobworms, crickets, larvæ of every kind, fall a prey in similar fashion to the different species of ants, *Formica*, *Myrmica*,

Lasius, Tetramorium, Tapinoma, etc. Even winged insects are not safe from them ; I have often seen butterflies, gnats, flies, and others which had entangled themselves in the turf, caught and killed by the ants."

Even comparatively large animals are not secure from their clutches, and are afraid of them. Moggridge, in his interesting work (London, 1873—74) on the " Harvesting Ants and Trap-door Spiders," of the borders of the Mediterranean, relates that the lizards are very fond of the winged males and females of the ants and follow their swarms, and that these are most gallantly shielded and defended by the working ants : " When, as often happens, the nest is placed in an old terrace-wall, one may see the lizards creeping along or lying moulded into the inequalities of the stones, all having their eager eyes directed towards the swarm. One may then see the worker ants walk with impunity straight up to the very noses of the lizards, while the male or female which should chance to struggle in the same direction would infallibly be eaten up. The lizards plainly show their fear of the workers by the way in which, when they make up their mind to try a dash at some outlying part of the ant-colony, they leap through the lines in the utmost haste, as if traversing a ring of fire. Now these worker ants are destitute of stings, and I can only suppose that their power of combination, stronger jaws and more horny coats, have gained them this immunity" (Supplement, p. 162).

What dangerous and dreaded creatures ants are in tropical countries will be shown later on.

Further, all these bodily and mental qualities vary as much or more in different races, species and individuals, as in different races and individuals amongst men. As Forel says, there is a greater difference between a *Plagiolepsis pygmæa* and a *Camponotus ligniperdus*, than between a mouse and a tiger ; and a colony of *Lasius fulginosus* compared with one of *Leptothorax tuberum*, is as Paris compared to a town or village. Strength, swiftness, the power of defence and attack, the number of the population of a colony, timidity, the time and frequency of swarming, smell, love of war, architectural skill, and choice of localities, the art of feeding the young, the habit of day or night work, and many other matters, vary, as Moggridge points out, between the widest limits. The bloodthirsty and murderous tiger-beetle, *Cicindela*, as is related by the same author, bears himself very differently when he chases the strong harvesting ants, and the small and weak *Formica erratica*. He seizes and swallows the latter without delay, while he is more than half afraid of the former. " I have seen this beetle lying in wait near a train of *structor* or *barbara* ants, watching until some individual separated a little from the main body, when it would rush forward and make a snap at it, retiring again as quickly as it came. If the tiger-beetle fails to seize its prey exactly behind the head it will let it go again, and two or three ants are often thus cruelly mutilated before a single one is carried off. No doubt the beetle has learned that if once this ant clasps its mandibles upon either antennæ or legs, nothing, not even death itself, will make it release its hold. It therefore tries to pin the ant in such a way that it cannot use its formidable jaws. Perhaps the habit of forming long compact trains may have been acquired by the ants partly with a view to guarding against attacks of this kind. The colonies of the little *F. erratica*, on the other hand, apparently have to trust to their habit of working under the covered ways which they construct, as well as to their activity and great numbers, for their preservation " (p. 164).

The most timid ant, according to Forel, of the species *Camponotus* is the *Camponotus marginatus*, which scarcely ventures to defend its nest, while the *Camponotus pubescens* is the strongest and bravest. The largest of the workers of the latter species compete with the famous Amazon, *Polyergus rufescens*. The harvesting ants also are not able to withstand the stronger *Camponotus*. Forel destroyed the partition-wall which divided a nest of the *Camponotus æthiops* from one of the *Atta structor*. The latter were so terrified that they fell a prey in the shortest time to the nippers of their opponents without serious resistance. The bravest species of ants are the above-mentioned Amazons and the bloodred ants, *Form. sanguinea*, of whose battles and robbery-excursions or slave hunts, exact details will be given later. Most of the species of the germs *Myrmica*, to which belong all the true harvesting ants, are dangerous and brave, while the comparative intelligence of the above-named bloodred ants appears to rise above that of all others. A very peaceful and gentle ant, which never ventures on a combat, is the *Botryomyrmex meridionalis*.

The English naturalist, Sir John Lubbock, who has made many careful attempts to investigate the character and life-habits of ants, and who has published his results in the Journal of the Linnæan Society (Zool. xii. and xiii. vols.) was able to establish a great distinction in the behavior of various individuals under exactly similar circumstances.

As there are in Europe alone more than thirty genera and more than a hundred species, and in the whole world more than a thousand species of ants, sprung from the same stock, it is easy to understand what a countless host of differences in bodily construction, character, intelligence, conduct, habits, manners, and so on, there must be—differences which it would take whole volumes to fully describe. We will only concern ourselves here with the most remarkable, striking, and best known species.

That the extreme intelligence of ants must be related to a special development of their nervous system and especially of their organ of thought, or brain, will be a matter of course to the anatomist and physiologist, who knows that the organ and the function—or its action under given circumstances—must co-exist side by side. For others, however, it is interesting and important to learn that the brain of the ant is comparatively the largest in the class Insecta, and is even more developed than that of the bee. According to a table compiled by Titus Graber ("On Insects," vol. i., p. 255) the volume—or size—of the brain of bees is the 200th, and that of the so-called accessory brain, or the "stalked bodies," [discovered by Dujardin, and named by him *corps pédoncules*. See *Annales des Sciences*, 3rd series, Zool. tom. xiv. 1850. p. 200.—Tr.] the 1000th part of the size of the whole body, of ants the 280th and the 600th part, while the brain of the cockchafer, which has no accessory brain, is only the 3000th part of its body. There is doubtless, therefore, the same difference as between men and the large mammals (horse, bull, etc.) which are subject to men, in consequence of their lesser brain and mental powers, although the latter are much smaller and weaker in body This is also true of the huge elephant, although its brain, corresponding with the size of its body, far exceeds that of man in development as to its bulk. None the less are the cerebral ganglia of the ant—which ganglia in invertebrate animals take the place of the brain proper to the vertebrate—no larger in reality than the quarter of the head of a pin, and the size again is of course different in the different species. " Under this point of view," says Darwin (" Descent of Man," ed. 1875, p. 54), " the brain of an ant is one of the most marvellous atoms of matter in the world, perhaps more so than the brain of a man." This fact shows at the same time " that there may be extraordinary mental activity with an extremely small absolute mass of nervous matter."

But the brain of the ant, not only in its relative size, but also in general shape and construction, is superior to that of all other insects, and most resembles that of hive-bees and the other socially-living insects of the same genus. At first sight we are startled to find two large projecting hemispheres, as in higher animals. If these are put aside, we find Dujardin's two " stalked bodies," surrounded by a cellular envelope, which gives to each hemisphere its rounded appearance. These stalked bodies are not so much developed in any other insect as in the ant, and are generally found to be either rudimentary or arrested. *Under* these is found the primitive ganglionic cerebral structure, which is common to all insects. It is a single oblique body, in the middle of a somewhat convoluted mass, and is covered by the lower part of the stalked bodies. Right and left lie the optic lobes, or centres of origin of the optic nerve; in front and beneath, the olfactory lobes.

The stalked bodies are not, as has been thought, centres for the so-called ocelli, or accessory eyes, but stand in a very decided relation to intelligence. They are very large in the branches of the genus *Formica*, which includes the most intelligent species; they are largest in the most intelligent of all the ants, the *Formica sanguinea*, or blood-red ants, and in the turf-ants, *Formica pratensis*. It is very remarkable also, that the neuter workers as far excel the winged males and females in the size and development of their stalked bodies, as they do in the amount of their intelligence. These bodies are the smallest in the unintelligent males.

It may be added that Treviranus had already proved that all the social Hymenoptera, to which genus belong bees, wasps, ants, etc., are distinguished from other insects by a more largely developed brain; but that Dujardin made the discovery that this larger development was due to the presence of these stalked bodies, which he described and which were named after him. He also proved that they are connected with intelligence, and almost or quite disappear in insects with poor intelligence. He found them very large in bees, comparatively larger yet in *Formica rufa*, the common wood-ant, to which the turf-ants belong as a sub-species.

The anatomy and physiology of the nervous system of all these intelligent insects, and especially of ants, needs, however, much more exact investigation than has as yet been bestowed upon them, and would doubtless bring to light many interesting particulars.*

Injuries to the brains of ants are followed by exactly similar results as injuries to those of higher animals, and the behavior of ants with injured brains is just like, or very much resembles, that of men or other mammals suffering from brain-lesions. First of all, any important brain-injury causes spasms and a number of involuntary reflex movements. Then succeeds a state of stupefaction, with an increase of reflex action, in which voluntary and conscious action fails. Thus an ant, whose brain has been perforated by the pointed mandibles of an Amazon, remains as though nailed in its place, a shudder runs from time to time through its body, and one of its legs is lifted at regular intervals.

---

* Since the above was written the author has become acquainted with an enquiry into the brain of ants, by Rabl-Rückhart, published in Reichert and Reymond's "Record of Anatomy, Physiology and Scientific Medicine," (1875, Vol. IV., p. 480,) which confirms and gives accurate details on the statements in the text. According to this writer, both the primitive lobes of the brain of the ant are covered with helmet-like or fungoid prominences, which Dujardin found in all social intelligent Hymenoptera, as bees, and which Rückhart distinguishes as lobes with "convolutions or radial striped circles." The convolutions he considers to be analogous with the convolutions of mammalia. In these prominences Rückhart also found certain "ring-like bodies" which consisted of peculiarly fine molecular matter. He compares the brain of the ant to a vertebrate brain perforated by the pharynx, which is of a strikingly higher type than the ganglia of the other Annulosa. A good description of the ant's brain, with its complicated anatomy, will be found in "Leydig's Tables of Comparative Anatomy," I., VIII, Fig. 4; and the same is reproduced by Titus Graber (Vol. I., p. 252).

It occasionally makes a short and quick step, as though driven by an unseen spring, but, like that of an automaton, aimless and objectless. If it is pulled, it makes a movement of avoidance, but falls back into its stupefied condition as soon as it is released. It is no longer capable of action consciously directed to a given object; it neither tries to escape, nor to attack, nor to go back to its home, nor to rejoin its companions, nor to walk away; it feels neither heat nor cold, it knows neither fear nor desire for food. It is merely an automatic and reflex machine, and is exactly similar to one of those pigeons, from which Flourens removed the hemispheres of the cerebrum. Just in the same way behaves the body of an ant from which the head has been taken away. In the numerous fights between Amazons and other ants, countless cases have been observed of slight injury to the brain, which have caused the most remarkable phænomena. Many of the wounded were seized with a mad rage, and flung themselves at everyone that came in their way, whether friend or foe. Others assumed an appearance of indifference, and walked serenely about in the midst of the fighting. Others exhibited a sudden failure of strength; but they still recognised their enemies, approached them, and tried to bite them in cold blood, in a way quite foreign to the behavior of healthy ants. They were also often observed to run round and round in a circle, the motion resembling the *manège*, or riding-school action of mammals, when one of the crura cerebri has been removed.

If an ant is cut in half through the thorax, so that the great nerve ganglia of the prothorax remain untouched, the behavior of the head shows that intelligence also remains untouched. Ants mutilated in this way try to go forwards with their two remaining legs, and beg with their antennæ for their companions' aid. If one of these latter lets itself be stopped, then we observe a lively interchange of thanks and sympathy expressed by the actively moving antennæ. Forel placed near to each other two such mutilated bodies of the *F. rufibarbis*. They conversed with each other in the above described way, and appeared each to beg for help. But when he put in some similarly mutilated ants of a hostile species, *F. sanguinea*, the picture was changed; war broke out between these cripples just in the same way and with the same fury as between perfect ants. Mutilated human soldiers have also been known to kill or try to wound each other, as they lay on the ground after an embittered struggle.

There is, moreover, a great difference in the behavior of ants which have only had the antennæ and eyes destroyed, and those which have suffered serious brain-lesion. The former show will and consciousness; the latter are automatic and their motions are reflex.

Their large and well-constructed brain would, however, be of small help or use to ants—as, for instance, the dolphin, living in the water and with a clumsy body, obtains little good from his large and well-developed brain—if a similarly well-developed and generally highly-organised body were not joined to it. They have in their feelers or antennæ—very long and flexible, consisting of a shaft and nine or ten joints, and thickened into a knob at the end, articulated with their heads by a very moveable hinge, and bent at an angle—an admirable organ for recognition and for mutual comprehension, which they thoroughly know how to employ. These are used for feeling with a continual rapid motion and touch. Within the antenna runs a very strong nerve, which takes origin at the base in a large ganglion. This nerve appears to supply not only the sense of taste, but also that of smell, so unusually developed in the ant. The antennæ are sensitive, even to the slightest movement. If they are cut off the mutilated ant loses the ability of finding its way, of distinguishing between friend and foe, and even of finding food which is placed near it. It does not even notice its much loved honey, unless its mouth accidentally comes into contact with it, and will put its fore feet into it, using these to feel with in default of its antennæ. It also tries to apply to it its whole head and the so-called mandibular palpi, but with very poor results. Ants whose feelers have been destroyed no longer trouble themselves about the building of their nests, the care of the larvæ, and so on. They generally remain quite quiet and motionless, and have as sad an appearance as men deprived of their principal sense.

The two fore legs are also very strong organs, and serve in building nests and in throwing up earth. Ants, off which Forel had cut both fore legs, made futile attempts at digging and building; they did not succeed in making a single respectable furrow. They also soon became covered with earth and dirt, as did also the larvæ and pupæ, for they were no longer able to clean either themselves or them. They nevertheless tried to look after their larvæ, but only succeeded in dirtying them, and finally left them to lie and perish. Ants have on their fore legs projecting ridges, which serve them as brushes and cleansing organs, and with which they constantly try to keep clean their heads, antennæ, mandibular palpi, mandibles, and thorax. The abdomen is cleansed by the other legs, which also have these ridges, only less developed. The legs clean each other in turn, and the spur is cleansed by drawing the leg backward and forwards between the mandibles and the mouth.

" It is very easy," says Forel, " to watch the Amazons when, returning from a marauding excursion, they go back to their homes over the surface of the ground slowly and wearied. We can then see how the fore leg of one side cleanses the antennæ of the same side, how it is then drawn through the mouth, and again rubs the antennæ. After awhile the same manœuvre is repeated on the other side. The Amazons clean themselves thus while on the march. They stop for a moment, while they hang on to a blade of grass by two feet on the same side, and brush with feverish haste the two hind legs of the opposite side and the abdomen with the spur of the corresponding fore leg. This only lasts some five or six seconds; they then recommence their march, soon to repeat the same task on the other side. They also from time to time cleanse their feelers."

This remarkable love of cleanliness is not, it seems, confined to the Amazons, but is common to all ants. The Rev. H. C. McCook, the distinguished American observer, found that the agricultural ants kept by him in captivity—which will be fully dealt with hereafter—were the cleanest creatures possible. The whole body was carefully cleansed after each meal or sleep, the different ants mutually assisting each other. The particulars of these interesting proceedings were exactly noted by the observer, and remind one of the fashion in which cats wash and rub themselves. A sense of comfort manifestly accompanies the action. The feelers, or antennæ, are cleansed with peculiar care. (Proceedings of the Academy of Natural Science of Philadelphia, 1878, 2nd April.)

Far less important than the antennæ are the mandibular

and maxillary palpi, which appear only to serve for seeking and perhaps also for testing food.

The next most important organs after the feelers are the generally toothed jawbones, nippers, upper jaws or mandibles, which give the ants their peculiar power and force, but which never serve, as was thought, for chewing or eating, but are only weapons and organs of prehension. Ants do not eat solid food, but only lick liquid or soft food with the tongue, like dogs. They tear or bite animals with their mandibles and then lick the soft interior. The mandibles are peculiarly developed and strong in slave-making species and in the so-called soldiers, which in some species are a separate caste, apart from the workers.

Of not less importance is the sting, situated in the abdomen, which is not found in all ants, but only in the genera *Myrmica* and *Ponera*. These sting severely, and pour into the wound a poisonous or irritating fluid from a poison-gland. Where no sting is present, the poison is dropped or spirted from the abdomen into the wounds made by the mandibles. Many kinds are able to spirt the contents of their poisoning glands at an assailant or enemy from a distance of several feet. The poison itself is the well known formic acid, which, as Taschenberg has observed, can be seen rising from the end of the abdomen of the enraged creature like a fine silvery fountain in the sunrays.

It is also worthy of notice that the whole alimentary canal of ants is divided into two parts, of which the front one serves the community, the hinder one the individual. The œsophagus is widened out into a kind of crop in the front part of that portion of it which lies in the abdomen, and this enlargement receives and can retain a great amount of liquid food. As soon as it is wanted, this food can be voluntarily regurgitated and given out, and serves as nourishment for the larvæ or for hungry comrades, for instance for males and fertile females which do not themselves seek for food, or, in some of the slave-making species, for the food of the lazy masters.

Thus the ant is thoroughly fitted for the important part it plays in nature and for the high place it takes in the world of insects and of appendiculate animals, not only by the high organisation of its brain and nervous system, but also by the total constitution of its unusually powerful and yet agile body, by the possession of well-developed organs of sense and powerful offensive and defensive weapons, of special instruments for building, digging, and cleansing purposes, and of an impetuous and fearless, but yet prudent and patient character.

CHAPTER V.

ANTS IN HISTORY.

THE remarkable qualities of the ant have at no time and at no place been quite hidden from man, and have been more or less recognised and praised by him. In some places in Arabia—as Freitag states in his Arabic-Latin Dictionary, vol. iv., p. 339, under the Arabic word for ant—they put an ant into the hand of the newborn child, so that the ant's virtues may pass into the young life. In ancient literature there are many notices of ants and of their remarkable characteristics. " Go to the ant, thou sluggard," says Solomon in his Proverbs, chap. vi., v. 6—8, " consider her ways and be wise; which, having no guide, overseer, or ruler, provideth her meat in the summer, and gathereth her food in the harvest." And chap. xxx., v. 25, "The ants are a people not strong, yet they prepare their meat in the summer." In the Mishna, the collected traditional and unwritten laws of the Jews, which was begun after the birth of Christ under the presidentship of Hillel, and preserves the memory of many old and otherwise forgotten habits and customs, the ants and their granaries are mentioned in connexion with the rights of the gleaners at harvest-time. "The granaries of the ants," it is written, "which are found in the midst of a growing cornfield, shall belong to the owner of the field; but of the granaries which are found after the reapers have passed, the upper half of each'shall belong to the poor, the lower to the owner." And it is further said: "Rabbi Meir is of opinion that the whole should belong to the poor, since in dubious points of gleaning, the doubt is always in the gleaner's favor." The object of the above cited law plainly is that corn gathered by the ants from the cornfields before harvest shall go to the owner, while that which is collected by the ants after reaping, and must therefore lie on the surface of their granaries, shall belong to the poor gleaners. Therefore the upper half is to belong to the poor, the lower to the owner of the field.

Among classic writers the Greek poet Hesiod, in his agricultural poem "Works and Days," speaks of the time whereat the prudent ant harvests the corn; and Horace, in his Satires (I. i. 1. 33), alludes to the foresight of the ant, which looks forward to and prepares for the future. Cicero ("De Nat. Deorum," lib. III., chap. ix.) says of the ants: "In formica non modo sensus sed etiam mens, ratio, memoria." (The ant has not only sensation, but also mind, reason, and memory.) Virgil, in the "Æneid," (book IV., 402) compares the Trojans, flying from Troy and laden with treasures, to ants laden with corn, carrying their booty home with eager haste. The Roman comic poet, Plautus, in his comedy "Trinummus," act ii., scene 4, makes a slave enter, who in order to make intelligible the rapid disappearance of some money entrusted to him, says: "It vanished in a moment as swiftly as a poppy-seed thrown to ants." Any one who has seen the precipitation with which certain Southern species of ants rush at grains of corn which are thrown in their way, will confirm the accuracy of this image.

Pliny ("Natural History," book ii.) says of ants: "You find among them a sort of Republic, as well as thought and prudence . . . . they hold meetings at which they mutually recognise each other. What proceedings go on! with what eagerness do they communicate with and question each other! . . . . We see how pebbles are worn out by their running over them, how posts sink deeper into the ground whereon they regularly go to work. A great example of the power of small but perpetual efforts."

Claudius Ælianus, who lived in the time of the Emperor Hadrian (221 A.D.) gives the following picture of the habits of ants, in his work on the "Nature of Animals" (ii., 25): "In the summer, when the ears are thrashed out after the harvest, swarms of ants come to the threshing-floors, in order to go thieving, sometimes singly, sometimes in bands. They pick out the grains of wheat or barley, and carry them off straight to their homes. Some busy themselves only with collecting, others carry away the load, and they understand very well how to get out of each others' way, especially the laden and unladen. These remarkable creatures, when they have arrived at home, and have filled their granaries with wheat and barley, perforate each grain in the middle. That which falls out serves as flour for the ants, and the remainder is incapable of germination. They perform this notable piece of domestic economy, lest the seeds, in the rainy season, should sprout, and so their food should be destroyed. So ants in this, as in other matters, have their fair share in the gifts of nature."

Ælianus further gives a very interesting account of the way of gathering and treating the corn, the particulars of which have mostly been fully corroborated by other observers. " When they set out on a foraging expedition, the largest ants march in front as generals. When they reach the harvest field, the younger ants remain at the foot of the grain-stalk, while the elder ones and the leaders climb up and throw down the gnawed-off grains to those waiting below. These then free the grain from the husk and carry off the seed. So ants obtain the food of man, which he ploughs and sows for himself, while they neither thresh nor winnow."

Ælianus appears also to have heard of the habit of ants in tropical countries; for he says (Book xvi., 15): "The Indian ant is certainly a clever creature. . . . . They leave an opening at the surface of their nests, through which they go in and out, when they bring the collected grain."

Aldrovandus, a writer of the seventeenth century, in his work on Insects (Book v.), speaks of ants, which store up grains of corn and gnaw off the radicles; but it is not certain whether he speaks from his own observation or only from hearsay.

The neatly told fable of Fontaine of the ant and the grasshopper is well known; it was borrowed by him from the old Greek fabulist, Æsop. Æsop relates : " The ants were busy one winter time drying in the sun the contents of their rain-drenched granaries. A grasshopper who saw this, and was at the point of dying of hunger, came near and begged for a morsel. Said one of the ants: 'What were you doing in the summer, you lazybones, that you need now beg for bread?' The grasshopper answered: 'I lived for pleasure. I sang and gave pleasure to the passers-by.' 'Oh, oh!' sneered the ant, turning away, 'dance in winter, if you sing in summer. Gather food for the future when you can, and think no more of playing and of pleasing the passers-by.'"

The ability of ants to understand each other was known to the ancients, as will have been seen by the above-cited passage from Pliny. The following anecdote is found in Plutarch ("De Solertiæ Animalium," chap. 11): "A certain Cleanthes relates that he has seen ants going from one ant-heap to the entrance of another, carrying a dead ant. Other ants came out of the latter, conversed with the newcomers and went back again. This scene was repeated twice or thrice, until at last a worm was brought out of the depths of the nest, which was evidently to serve as a ransom for the dead body. Then the ants which had brought the corpse, left it lying there, and carried away the worm instead."

However incredible this may sound, it is beyond doubt that ants and bees have been seen carrying away and even burying their dead, and of this further details will be given later.

From this same passage of Plutarch we learn further that the ancients were acquainted with the habit of ants of biting off the germinating radicles of seeds in order to prevent their growth, and that they had also seen ants, carrying loads which were too heavy for them, try to lighten them by biting off pieces.

That which most attracted observation in ancient times was the habit of ants living in the South, of gathering corn and laying up stores for the winter. Of their other remarkable characteristics, such as their wonderful social polity, nothing or very little appears to have been known, while attention was directed more to the study of the animals whose size more readily attracted notice. When an animal is very minute, people are apt to think that its organisation must be very simple, and its intelligence very small, and the influence of this prejudice over the majority is great. The gigantic dimensions of a whale, or of a reptile of the fossil age, attract general attention, while this attention is difficult to arouse, if the most wonderful phænomena are exhibited in the life of a gnat or an ant. And yet the extraordinary capabilities of an apparently lowly creature are exactly that which yields to the philosopher the most valuable results. But the incredible fineness of sense in insects ought to have warned the observer that corresponding intellectual capabilities would certainly be present. For of what use would such senses be to a creature which was unable to utilise them on account of its poor intellectual organisation? Or for what object would insects, and especially ants, possess such great muscular and bodily strength, which renders them capable of performances which comparatively surpass twenty, thirty, and even a hundredfold those of men or of the larger animals?

Even in the last century Hermann Samuel Reimarus, the then most distinguished writer on animal intelligence, and one who in many ways is unsurpassed to-day, says practically nothing about ants in his famous book on the skill of animals (Hamburg, 1762). He merely mentions in cursory fashion in section 121, a "good story of an Ant-Republic destroyed by him" related by Professor Meyer ("Attempt at a new science of the Intelligence of Animals," Halle, 1750), while he appears to throw some doubt on the consequences thence deduced by Meyer. But it is at least evident from this that something was even then known as to the peculiar character of the social polity of ants, and that it was designated a "Republic." Ants do in reality live in a Republic, in the fullest sense of the word, that is, in a state "on the widest democratic foundations," as people used to say in 1848; and it is very noteworthy and significant that just the most intelligent family among socially-living insects enjoys and has made for itself a polity which is regarded among men as the relatively best and most ideal, while a step lower, among bees, we recognise a distinct inclination to the form of so-called constitutional monarchy. Among men it is usual to say that the republican form of government, from the theoretical standpoint, best represents the ideal of the state and the principles of justice, as well as of universal equality; but that having regard to the ineradicable weakness of human nature and the consequent impossibility of self-government, it is not practically realisable. If this be true, we men have certainly no ground from which to look down contemptuously upon the little ant-nation, every tribe of which considers itself intelligent and civilised enough to be able to live under the principles of universal equality and liberty.

But, as though this were not enough, the Ant Republic is not only a political, but also a social or socialistic Republic, and in reality is that which our most idealist political and social reformers are wont to put forward as the last and mightiest aim of human efforts after perfection in this direction, that which Plato and Thomas More have already pointed out as such. If the democrats of the latest style had to find a polity fashioned after their own ideas, or to govern a "Proletariat State," no better advice could be given them than to take as model the political and social arrangements of ants. The ant-state is a "Proletariat State" in the fullest and truest sense of the word, in the direction of which only the wingless, sexless, worker-ants, which have no families of their own to look after, take part, while the winged males and fertile females are kept as prisoners in the nest, and are only fed and nurtured for the sake of their progeny. The expression "sexless" is really not appropriate to the men, or rather women-workers, for these are really females with arrested sexual organs, so that the state is constituted under completely feminine rule. As P. Huber remarks, they are women whose moral qualities have been developed at the cost of their physical. The individual ant does not possess a family, for the principle of public and state training of children—such as the philosopher Plato is known to have desired in his Republic, and which indeed would be necessary in a fully-organised "Proletariat State"—is thoroughly carried out in the Ant Republic.

That, further, the worker-ants were formerly winged, and were therefore more like the perfectly developed females than they now are, is shown in the interesting discovery, lately made by Dr. Dewitz, of rudimentary or arrested wings in the pupæ.

The males and fertile females, which (the former more than the latter) are far behind the workers in intelligence, and only live for amusement and for propagation, are, as we have said, kept in the nest like prisoners, and have no other vocation than the maintenance of the colony; even this vocation is carried on only by the permission of the working population and under their strict superintendence. On a warm sunny day they are allowed to quit the nest, or dwelling, and for a change to take a constitutional on the surface. But they are then watched by a number of workers, which prevent them from flying away. There is moreover this great difference between the males and fertile females, that the males are far more behind the workers in intelligence and other good qualities than are the females. Owing to the weakness of their mandibles they are as incapable of every kind of work as they are of defending themselves against an enemy. Forel thinks that they are not even able to distinguish between the workers of their own colony and those of a hostile tribe. If a nest be destroyed, they try to hide away in corners, and often do not know the way back to their tunnels, while the fertile females are quite able to find it. The latter will have long found a place of safety, while a number of males are wandering unsteadily about, without knowing whither to flee. The workers often find themselves obliged to lead them all back to the nest.

The fertile females are also able to occasionally help the workers in their labors, a task of which the much smaller males are quite incapable, and Forel has often watched these females carrying larvæ and pupæ. They manage to follow the fleeing workers, in case of flight, while the males are scarcely ever able to do so. They also exhibit many signs of courage and intelligence, although they never equal the workers, and they show in their character a certain irritability and want of perseverance. They are also able to distinguish between friend and foe. They are almost twice as large as the males, and are generally rather larger than the worker-ants.

It is perceived that among ants the female sex has predominance over the male in all questions of rights, and this is carried out in a way which must raise the envy of the boldest champions of human female emancipation.

When the fertile females go on the surface of the nest in company with their unworthy wooers, or future husbands, they are, as we said above, accompanied by a guard of workers, which appear to be very proud of the importance of their task, and lead their charges back within the nest at the least sign of danger. After these walks have been repeated during several days, the great flight, or wedding tour, takes place with the consent of the workers, generally on some fine afternoon in July or August. The entrances to the nest are widened or increased in number, for the greater convenience of the swarmers, and an unusual amount of commotion is seen on the surface of the nest. One of the fertile females begins to flutter her wings and to rise in the air: a second follows; the males do the same in order to follow their wives. The watching workers who, having no wings, are unable to follow, become much agitated; at last the interesting scene closes with the flight of the whole swarm into the air, as a thick cloud, rising to a considerable height.

For the flight is chosen a fine, bright summer day, as when a storm has taken place the day before and there is no other to be feared; and since such a day is not only chosen by one but by many colonies in the same neighborhood, there are often swarms so large and so extended that they darken the air, and are mistaken for clouds of smoke arising from a distant fire. These swarms dance gaily up and down in the air for hours, generally circling round some high place, such as a church tower, the top of a tall tree, a hill etc., and the fertilisation of the females by the males is effected during the flight.

Zoologists have not yet quite ascertained if the merry bridal folk during their happy journey sing the well-known song:

"Das flüchtige Leben eilt schneller dahin,
Als Räder am Wagen,
Wer weisz, ob ich morgen am Leben noch bin?"

But in any case such a song would be eminently suitable to the occasion, for a sad end soon comes to the brief happiness. When the wedding parties return to earth after a few hours' flight, the poor husbands perish in shameful wise, for they are unable to feed themselves, and either die from hunger or fall a prey to pursuing birds and spiders. Many also are killed by hostile ants. The workers or neuter-ants of their own colony have lost all interest in them from the moment of their return, and trouble themselves no more about them, for they well know that the males have now fulfilled their vocation, and that their prolonged existence as useless consumers would only be a burden. A very egoistical, but also a very republican, or rather socialistic proceeding. For "He who will not work, neither shall he eat." We shall see later that this beautiful precept is yet more severely and cruelly obeyed among bees.

The greater part of the remaining females have no happier

fate than their husbands, for their number is far too large for work to be found for them all. A few who chance to find a hole or to dig a refuge in moist earth at the spot whereon they descend, become the mothers of future colonies. For the performance of this task the wings necessary for the wedding tour are only burdensome and hindering; and a wonderful instinct—if people will insist on the expression—teaches them to voluntarily free themselves from an organ thenceforward useless and even mischievous. They therefore scratch at their wings, one after the other, with the clawed ends of their feet, and drag and twist them about until they fall off. This is the easier, because the articulations of the wings with the body are very slight. The operation, further, does not appear to give them any pain. When this business is over, the fertilised females are queens, and they really have a kind of kingdom, for they are compelled by passing ants to return to their native home, and are there kept for the important business of egg-laying.

So at least are things managed according to the majority of writers on ant life; but Forel entirely contradicts this account, and maintains that such females only become queens of the nest as have not shared in the swarming or wedding-flight, and which have been fertilised within or on the surface of the nest. According to him the females which take part in the swarming *never* return to their own nest, and, indeed, show a dislike to it. Therefore the workers keep back a number of fertilised females, which have been fertilised on the surface of the nest or close by before the wedding-flight, and use these as the mothers of the colony. They thereby ensure the important point of preserving pure blood in the colony, since foreign males never venture on the surface of another nest, while during the swarming in the air there is great and unavoidable mingling of the members of different colonies. The females thus kept back quickly accustom themselves to captivity and do not seek to escape. Sometimes there are only a few of them, at other times twenty, thirty, or even more. They are saved the trouble of pulling off their wings, for the workers look to this business for them, and tear or bite them off.

Forel's view must command credence when we remember that the wedding-flights often go very far from their nests, and that owing to the mixture of the various swarms it seems almost impossible that each female should find her way back to her own nest on returning to earth: and the more so that the behavior of the workers described by him manifests the attainment of a defined and very important object. They have all they want in the queens which have become such under their direct oversight and co-operation, and they need no others. These are cared for, cleaned, brushed, nourished, and nursed through their egg-laying, in the best fashion.*

Each queen has generally a suite of ten or more workers, which are constantly busied about her, and take every imaginable care of her and her eggs. But this is not the case always nor with all species. Forel noticed that in the genus *Leptothorax*, the queens live like the workers, but are less skilful at work. Other species give their queens the best and largest cell in the nest, and are so attached even to her corpse that it is difficult to tear them away from it. The greatest care for the queens was observed by Forel to be bestowed upon them by the *Lasii*. Here they are always surrounded by a number of workers, which follow them everywhere, and often so cover them with their bodies as to render them completely invisible, while they also feed them and gather the eggs laid by them. The queens appear to exist for the sake of the workers, so that they are wrongly called "queens," and, indeed, in a Republic this could not be otherwise. Their queenly dignity appears to consist less in command than in obedience, and the care and attention which is showered upon them is less for their own sake than for that of their young.

---

* A very well informed critic of the Dutch translation of this work, writing in the "Groniger Deekblad" of June 16, 1877, considers that the above views of Forel are contradicted by the fact that in such case the described swarming would be objectless. He thinks, on the contrary, having regard to the injurious effects of in-breeding, that it is of the greatest importance to the ants "not to keep the colony pure, but to have an infusion of fresh blood." He regards the fertile females which have not left the colony, and which, according to him, are only fertilised *after* the bridal flight, as a kind of reserve corps to which the workers resort only in case of need and if they fail to secure any returning queens. They are only the exceptions, while the swarming females are the rule. The Termites, according to him, (and this will have exact mention later,) have such a reserve corps.

Henceforth the care of these young is the chief business and the special duty of life of the workers, and it is very worthy of observation that this social tendency is most strongly developed exactly in those creatures which, on account of their sexlessness, are not in a position to have a family of their own and to propagate the society. Every care is bestowed on the eggs laid by the queen, which, according to the observations of Huber, grow in a wonderful and almost inexplicable way, before the larvæ, or grubs, are hatched. Since the workers gather the eggs together in little heaps, and lick them constantly with their tongues, the growth apparently takes place by endosmosis; *i.e.*, through the passage inwards of nourishing matter from without. If the workers are removed from the eggs the latter perish from dryness, which proves that the licking is necessary for the preservation of the life of the egg. After about fourteen days, during which the workers sometimes place the eggs in the higher and sometimes in the lower stories of the nest, in order to protect them from too great heat, cold or damp, small, white, almost inactive little grubs emerge, maggots or larvæ without eyes or feet, which can only exist through the nursing of the workers, and whose care requires a far higher kind of attention than the eggs. They are thoroughly well nursed, and, as Blanchard says, no more attentive, watchful nurses, or more devoted to their duty, can be found than these. The larvæ can scarcely move and cannot change from place to place. Still less can they eat alone, so that they are thrown entirely on their nurses for support, and these feed them in the same way as birds do their young. The eagerness with which ants are known to seek nourishment, especially sweet liquids, is due far less to their own needs than to those of their foster-children. They feed these just in the same way as they do their companions and queens, regurgitating into the mouth the food stored up in the crop or proventriculus, and then giving it from mouth to mouth. The larvæ have nothing to do save to take with their tongue the nourishment offered to them, and they show their hunger by stretching out their little brown and generally somewhat retracted heads. During feeding they always lie on their backs; and since the young ants, after emergence, as will be later shown, are fed for some little time by the older ones, it is easy to see how much tendance, care and work are requisite for the complete rearing of each individual ant.

The workers lick the larvæ almost ceaselessly, clean them if they get soiled with earth, and, when needful, carry them about as they do the eggs, to the different parts of the nest. The larvæ are also occasionally taken out in distinct groups, separated by the older and larger, so that one is involuntarily reminded of a school, with divided classes of elders.

"Nothing is more interesting," says Blanchard, "than to watch the ceaseless care of the ants for their larvæ. They clean them by rubbing or brushing them with their mandibular palpi; in the mornings they carry them to the upper parts of the nests, in order that they may enjoy the pleasant warmth, while, later in the day, they shelter them in the lower rooms from the burning rays of the sun. These carryings up and down are repeated as often as is required by atmospheric changes. It is surprising to see the tender care with which the ants carry the soft and easily injured bodies of the larvæ between their hard horny jaws. An accident never happens to them, they are never crushed, nor wounded, nor knocked against the hard walls of the long tunnels of the nest."

When the larvæ are grown, which is during the summer or in the following spring, they spin themselves a cocoon for the change into regular ants, and are then called pupæ, nymphæ, or popularly, "ant-eggs." Their egg-like appearance and clear smooth outside make most people incorrectly regard them as the real eggs of the ant. They are very much sought after, as they are the chief ingredient in the food of caged warblers.

The nymphæ need no food, but are carried about like the larvæ by the workers, licked, cleaned, and exposed in heaps in front of the nest on fine days to the heat and light. When the sunrays fall on the nest, the ants outside, the sentries, give a signal to those within, and the workers inside quickly carry the pupæ and larvæ on to the surface of the nest, and later carry them in again, first to the upper story, where it is still warm. They pull along the large white shapeless things, holding them in their strong jaws, "like cats their kittens," and try to defend them at the sacrifice of their own lives against the marauders from foreign ant-colonies—as will be hereafter described—or to save them by flight. If, for instance, we break a hole in an ant-hill, we at first see the larvæ and pupæ in the upper floors rolling in heaps over each other. The workers who are in the neighborhood at once rush to help. Each seizes a larva or nympha and carries it away. A moment later, in answer to a signal of alarm, hundreds of the sterile ants come up from the depths of the nest, and throw themselves upon the remaining larvæ and pupæ in order to hide them in the lower rooms of the dwelling. As soon as this first and chief duty is accomplished, they at once turn to the repair of the mischief which has been done, and this with such speed and ability that generally in about an hour there is no sign of the destruction left.

Although the pupæ want no food, they yet seem unable to exist without the tendance of the ants; at least Forel was unable to keep them alive without this help. The aid of the workers is quite indispensable, when the great moment draws nigh, in which they are to break from the cocoon, their envelope, to enter upon the life of the true ant. The pupæ are generally unable to free themselves from their covering—although some observers deny this —and would perish if left unassisted.* They do, indeed try to release themselves, but do not generally succeed, and are only able to set free a part of their bodies. The workers, therefore, open the cocoon, or case of the pupa, with their sharp jaws, and draw the young animal out carefully. At other times they only help to free the legs and wings. The workers do not keep to any exact time for the ripening of the nymphæ, but free them sometimes earlier, sometimes later, " as is most convenient," as Forel expresses it when mentioning this observation—a fact which does not suggest that the workers are here guided by never-erring instinct. The pupæ would of course perish, if they were freed either too early or too late.

The little animal when freed from its chrysalis is still covered with a thin skin, like a little shirt, which has to be pulled off. When we see how neatly and gently this is done, and how the young creature is then washed, brushed,

---

* Mr. McCook found that pupæ which he kept away from the workers always perished, whence he concluded that the assistance of the latter is necessary. — " Transactions of the Am. Entom. Soc.," Dec., 1876.

and fed, we are involuntarily reminded of the nursing of human babies. The empty cases, or cocoons, are carried outside the nest, and may be seen heaped together there for a long time. Some species carry them far away from the nest, or turn them into building-materials for the dwelling.

The pupæ, entirely freed from their cases, and now young ants, are however not yet ready and armed with all the advantages of a true ant—as they ought to be if the instinct-theory were true—but still need the help and guidance of their elder sisters. As already mentioned, these at first give it food, and begin to lead it about the nest, and train it in a knowledge of domestic duties, especially in the care of the larvæ. Later, and after their originally soft skin has become hard, they are taught to distinguish between friend and foe—to fight, whereby the important consciousness that they belong to a certain race or a certain colony is awakened in them. If a colony is besieged or attacked by another, it will be noticed that the young ants, which can be easily recognised by their lighter color, never mix in the fray; they only understand enough to follow their friends in flight, or to drag away the pupæ.

In order to establish this remarkable fact, Forel undertook the following experiment. He brought young ants from three different species into a glass case with pupæ of six different species; all the species were more or less hostile to each other. Damp earth with a piece of glass over it represented the nursery. The young ants set to work together without quarrelling, and carried the pupæ under the glass and established them therein. One individual, which was somewhat older and darker, belonging to the species *rufibarbis*, separated itself and carried its pupæ on one side. Forel several times tried to make it approach the others; it turned back to its corner each time, until at last, as it could not get out of the way of its enemies, it decided to unite with them. When the pupæ were ready for emergence, they helped the young ants by tearing their cocoons without respect of persons, although Forel noticed that the members of each species very willingly had to do with the pupæ of their own species. Only the pupæ of the *F. æthiops* were somewhat neglected, so that they perished in their closed cocoons. Thus Forel reared an artificial ant-colony out of five different species, which lived together on the best terms.

Ten days later, when the young ants had become darker, Forel took his colony and installed it in a chink in a wall in the open. In order to strengthen it, he brought some young workers of the same species from a foreign colony. But the original ants would not receive them; they first threatened them with their jaws, and then seized them and carried them a long way off and left them. They repeated this several times. The original and artificially united ants thus formed themselves into an independent colony.

The education of the young ants by the old goes on tolerably rapidly, and Forel saw that in several species the young were able to distinguish their friends more or less at the end of three or four days. Owing to its rapidity many observers have entirely overlooked this education, and have thought that ants come into the world with their whole knowledge and mind ready and fixed. Still more important is the teaching of building, of which, later, exact account will be given.

The natural superiority of the older ants over the younger, given by age, strength, and experience, appears to be the only personal inequality which is to be found in this Republic of Liberty and Equality. The most trustworthy observers unanimously agree with the sentence quoted from Solomon, that the ants, like the colonies of bees, wasps, etc., have no chiefs, overseers, nor leaders, and that one is as good as another. The consciousness of their duty alone keeps them orderly and at work. Some observers, as for example, Ebrard, have nevertheless spoken of such overseers. But Forel affirms that they are the creation of his own fancy. Huber, he goes on, has already shown that ants have no overseers, and that even the slaves never suffer the least oppression at their masters' hands. He himself confirms this, and never saw an ant which played a submissive part towards its companions. A worker of larger size is always an object of greater attention from the others than is a small, but only on account of its size; if the large ones march at the head in their excursions, it is for defensive purpose, of which the smaller are not so capable. On an occasion of change of dwellings, there is no difference of industry among the different ants. The smaller are only more busy because the larger are more warlike, and more fit to do battle. The soldiers, or fighters, which among some European and most tropical ants, are, as has already been mentioned, a special class distinct from their comrades, of whom more will be said later, never play a commanding *rôle*, but only one of service to the community. McCook (*ante*. Nov. 1877, p. 178) establishes that each ant develops a complete personal independence, and that there can be no talk of overseers, of leaders, nor of rulers. May we not naturally ask why so extended a self-government, which is enough for these little Republicans, is not also possible for man? And why should we take it for granted that in a perfectly free community men would only work if compelled, when these animals give proof that such a free commonwealth is very possible, and is compatible with the voluntary work of all?

As to the so-called queens, it has been already said that they do not exercise the smallest authority, and only so far deserve their name in that, as a rule, they take no part in the ordinary work, and that, omitting their duty of egg-laying, they yield themselves up to a *dolce far niente*, a soft idleness, to a frivolous and careless life of pleasure. They also resemble human queens in allowing themselves to be fed by their *quasi*-subjects, but they are very favorably distinguished from their human antetypes, so that under exceptional circumstances, when there is need, they set to work and are not ashamed to perform the same tasks as their subjects. This is especially the case where there is a lack of workers' hands. Lespès gives a most striking example of this kind. He watched a small species of ant in Southern France, which gathered in very small colonies, generally consisting of only about sixty members, and among these not less than twenty were queens. Here the latter shared the work. Such an arrangement seems very foolish, and does not say much for the famous "design" in natural order. But the misarrangement is again compensated for, since the queens, as we have said, willingly resign their privilege of idleness, and, forgetting their queenly dignity, share the work of their subjects. Has anything like this ever been heard among men? Even among these are found political abnormalities reminding us of the misarrangement among the ants. Let us think of the negro princelings in Africa, or of the constitution of our former German Empire, in which some hundreds of sovereign or independent princes, counts, bishops, archbishops, etc., ruled a few millions of subjects.

Or of the beautiful times of romantic chivalry, when every spurred noble commanded a large or small number of bondmen or personal retainers. But have we ever heard of any one of these little lords rising to the self-renunciation of the royal ant-mind, and taking part in the useful work of his subjects? Truly, man ought not to forget that only men have reason, and that animals, according to the philosophers, have merely instinct!

CHAPTER VI.

Ant Architecture.

BUT to return to our subject and our busy worker-ants, whose tireless and well-directed industry is not confined to the high task of training their successors, but extends over a whole circle of other scarcely less important duties.

Before all, there is the very complicated dwelling, which is not seldom built, partly in, partly above the ground, of from twenty to forty and more stories one over the other, thus leaving far behind it the highest building of man. This so-called ant-heap—the outer appearance of which does not in the least betray the wonderful complexity and appropriateness of its inner arrangement—often rises over a metre in height above the ground, while it also stretches as far below the surface, making altogether a height of between two and three metres. It is built in most artistic fashion out of all attainable materials, such as wood, earth, stones, leaves, stalks, twigs, etc., anything that will serve the ants' purpose. On commencing the building, the earth is first hollowed out in a given circle; then the building itself is made by help of the thrown-up earth and the other materials brought there, and is widened and enlarged later according to necessity and the condition of affairs. Each story is supported by shafts, or pillars, or beams of wood or clay, the steadiness of which is subjected in each case to careful testing. These beams are often of remarkable length and strength. Forel found some which were thirteen centimetres long by one and a half millimetres in diameter, and others of five centimetres long by three and a half millimetres deep. There are some even larger, and a single ant is able to drag them. Their great length allows the ants to make large rooms or halls within the dwelling. This is specially wanted for the middle of the so-called labyrinth, which consists of a large chamber, the roof of which is held up by a scaffolding of beams laid crosswise. Round this are numberless single rooms and galleries, which are divided by partition-walls, and arranged in stories. The lowest parts of the nest are built most firmly and methodically. It is highly interesting to see how the ants drag the beams along, and overcome all the obstacles in their way. Two or more workers usually unite in consultation over the arrangement of the work, when a little way has been made. One generally begins in one direction, one in another. Or both work in the same direction; but a grass-stalk, a plant-stem, or something of that kind gets in the way and renders their endeavors abortive. They soon recognise their mistake, and unite their efforts in a more fruitful direction.

"Looking at the entangled mass of sticks of almost the same size," says Blanchard, "which the ants have heaped together, it might be thought that they were the work of chance. But a more accurate investigation convinces one that these bits of wood are arranged with wonderful nicety, that rooms, galleries, lodges, tunnels, etc., are all so made as to yield easy communication between the several parts of the nest. The spaces between the beams are filled up with earth, pebbles, dry leaves, etc., the chinks are plastered over, the roughnesses smoothed; pillars and pilasters are reared of damp earth; in brief, the creatures act in everything like clever architects."

Clever as these little architects may be, they are, however, subject to error, like human architects, and have to suffer from the clumsiness of some of the laborers. Yet, on the other hand, they do not find it difficult to repair blunders which have thereby happened. Wrongly directed walls are pulled down and others built: workers which have done bad work are corrected by the others, or are put right and obliged to work under the supervision of a companion. P. Huber relates: "After I had observed the ingenuity with which these ant-palaces were built, I felt that the only way of finding out the true secrets of their construction, was to watch the behavior of the several ants at work. My daybooks are full of similar observations; I will quote some of those which appear to me to be of sufficient importance. I will picture the conduct of a single ant, which I was able to follow for a long time. I once saw an ant scraping up the earth near a hole which served as entrance to an ant-nest.

It heaped together the stalks it had loosened, and made little beams of them, which it carried there and then into the nest. It always came back to the same place, and appeared to follow a settled plan: for it worked with eagerness and patience. I then discovered a lightly traced furrow in the ground. It ran in straight lines, and appeared to suggest the beginning of a passage or gallery. The worker, all whose actions went on under my eyes, deepened and widened the furrow, and smoothed its sides: and I was at last convinced that its aim was to make a way from a certain hole, to the entrance of the underground nest. This passage, made by a single worker, was from two to three inches long, and was open above and bounded on each side by a wall of earth. Its shape, resembling that of a gutter, was perfectly regular; and the little worker had not left therein a grain too much. The work of this ant was so superior and accurate that I could almost always recognise what it wished to do and what part it built. Near the opening, where the passage ended, I found a second ant, making a road exactly similar. Between these roads ran a little wall, three or four lines in height. When the ants are busied in building together a wall, a room, or a gallery, it sometimes happens that the different parts do not perfectly agree. This is not infrequent, but the ants do not allow themselves to be thereby disconcerted. I will relate one case, wherein a worker discovered its mistake and knew how to remedy it. A wall had been partly erected, which appeared as though it were intended to support the still unfinished arched roof of a large room, which was being built from the opposite side. But the worker which had begun the arch, had given it too low an elevation for the wall on which it was to rest, and if it had been continued on the same lines it would have met the partition wall half-way up, and this was to be avoided. I had just made this criticism to myself, when a new arrival, after looking at the work, came to the same conclusion. For it began at once to destroy what had been done, to heighten the wall on which it was supported, and to make a new arch with the materials of the old one under my very eyes. When the ants begin an undertaking, it seems exactly as if an idea slowly ripened into execution in their minds. Thus, if one of them finds two stalks lying crosswise on the nest, which make possible the formation of a room, or some little rafters which suggest the walls and the corners, it first observes the various parts accurately, and then quickly and neatly heaps little pellets of earth in the interspaces and alongside the stalks. It brings from every side materials that seem appropriate, and sometimes takes such from the uncompleted works of its companions, so much is it urged on by the idea which it has once conceived, and by the desire to execute it. It goes and comes and turns back again, until its plan is recognisable by the others." Ebrard (" Etudes de Mœurs," Genève, 1874, p. 3) relates:—

" The earth was damp and the workers were in full swing. It was a constant coming and going of ants, coming forth from their underground dwelling, and carrying back little pellets of earth for building. In order to concentrate my attention I fixed my gaze on the largest of the rooms which were being built, wherein several ants were busy. The work had made considerable progress; but although a projection could be plainly seen along the upper edge of the wall, there remained an interspace of about twelve or fifteen millimetres to fill in. Here would have been the place, in order to support the earth still to be brought in, to have had recourse to those pillars, buttresses or fragments of dried leaves, which many ants are wont to use in building. But the use of this expedient is not customary with the ants I was observing (*F. fusca*, dark grey ants). Our ants, however, were sufficient for the occasion. For a moment they seemed inclined to leave their work, but soon turned instead to a grass-plant growing near, the long narrow leaves of which ran close together. They chose the nearest, and weighted its distal end with damp earth, until its apex just bent down to the space to be covered. Unfortunately the bend was too close to the extremity and it threatened to break. To prevent this misfortune, the ants gnawed at the base of the leaf until it bent along its whole length and covered the space required. But as this did not seem to be quite enough, they heaped damp earth between the base of the plant and that of the leaf, until the latter was sufficiently bent. After they had thus attained their object, they heaped on the buttressing leaf the materials required for building the arched roof."

One of the most industrious species is the black or black-brown garden ant, *Lasius niger.* These build in stories, but, according to Huber, follow no distinct plan. It rather appears as though nature left them a certain elbow-room, and that they can mould their original plan to suit circumstances. The larger spaces, or rooms, are always supported by smaller rooms and sometimes by regular arches. Some rooms have only a single entrance or a communication with the chamber beneath ; other very large ones are perforated on all sides, and resemble a centre towards which all roads converge. On opening such a nest, the rooms and large spaces are found filled with adult ants, while the chrysalides or pupæ are brought into the rooms nearer the surface according to the time of day and the temperature. For, as Huber remarked, ants are gifted with a very fine sensibility in this respect, and know exactly the degree of warmth which is suitable to their young. The stories being built one over another, it is easy to find each desired modification, while during floods caused by heavy rain the whole community takes flight from the upper to the lower parts of the nest.

Huber describes as follows the exact fashion of building :
" Each ant carried in its jaws a little ball of earth, which it made while it scraped the floor of the lower stories with the ends of its jaws or mandibles. These balls were again divided by the toothed mandibles and set beside and above each other, and the whole was fastened and lightly pressed together by the forefeet. Each movement was followed by the feelers, each bit of earth tested by them. The whole work went forward very rapidly. After the ground-plan had been sketched, and the foundation of the future pillars and walls marked here and there, the further building followed. Often two little walls, designed to bound a gallery, were observed being built at a little distance from each other. When they had risen to a height of from four to five lines, an arched roof was begun, in order to unite them, by making a projection of damp earth along the upper edge of each wall, and this was increased until each met that of the opposite side. The galleries made in this fashion were often a quarter of an inch wide.

" Here several perpendicular walls marked the entrance to a room which was designed to communicate with different corridors. There was a shapely saloon, supported by numerous pillars. Further on was recognisable the outline of one of those central spaces, which were spoken of above, whereon many passages opened. These spaces were the longest; but although they sometimes had a width of two and more inches, the ants had no difficulty in roofing them in. They began the work in the corners wherein the walls met, as well as along the upper edges of the latter; from the top of each pillar, as from so many centres, bow-like arched layers of earth were built horizontally, so as to join those which sprang from the other sides of the large open space to meet them. The number of little wall-builders coming from all sides with their bit of mortar in their mouths, the regularity they observed, the unanimity with which they worked, the skill with which they took advantage of a shower of rain so as to enlarge their dwelling—all this offered a most interesting spectacle to a lover of nature.

" I have sometimes feared lest their building should be unable to support its own weight and a heavy shower of rain. But I was reassured when I noticed how firmly the different parts were fitted together, and that the rain could only fasten them together more permanently. Only a very heavy rain occasionally breaks down badly built rooms, but they are very quickly repaired.

" A complete story is often built in from seven to eight hours, and the several ceilings all form at last a common platform. As soon as such a story is finished, they generally begin to build another on its surface, particularly when the weather is favorable. A dry wind stops the work. I noticed how a strong north wind quickly dried the new buildings, and made them fall to pieces. When the ants are convinced that their work is useless, they do not cease building, but destroy all unfinished rooms and walls, and spread out the *debris* on the surface of the top story."

Forel confirms all these observations, adding the remark that it is rare to find a completely uninjured nest standing alone, so as to be able to recognise all these arrangements. According to him also, the nests of *F. fusca* show great irregularity, because each ant works independently of the others. Every sort of material—earth, grass-stalks, snail-shells, leaves, roots—and any outside support suits them to build with or upon. On the great dome of the nest many smaller domes or towers may often be observed, which however generally disappear in the autumn.

Especially must it not be thought that all ant nests are constructed on the same or on a similar plan, as would of course be the case if the little architects followed an innate instinct. On the contrary, the greatest diversity is found according to circumstances, seasons, genera, and species. Even the same species builds very differently under differences of place, climate, or other circumstances. Thus, for example, the *Lasius acervorum* never builds under stones in plains, while in the Alps it shows a decided inclination to build under the same stones as the species *Myrmica*. This circumstance alone—omitting others—would suffice to set the intelligence of ants above those of bees, which are known to build their cells everywhere on the same almost unchangeable model.

" The characteristic *trait* of the building of ants," says Forel, "is the almost complete absence of an unchangeable model, peculiar to each species, such as is found in wasps, bees, and others. The ants know how to suit their indeed little perfect work to circumstances, and to take advantage of each situation. Besides each works for itself and on a given plan, and is only occasionally aided by others when these understand its plan. Naturally many collisions occur, and some destroy that which others have made. This also gives the key to understanding the labyrinth of the dwelling. For the rest, it is always those workers which have discovered the most advantageous method, or which have shown the most patience, which win over to their plan the majority of their comrades and at last the whole colony, although not without many fights for supremacy. But if one succeeds in obtaining a second to follow it, and this second draws the others after it, the first is soon lost again in the crowd."

Espinas also observed ("Animal Communities," German ed. 1879, p. 371) that each single ant made its own plan and followed it until a comrade, which had caught the idea, joined it, and then they worked together in the execution of the same plan.

This observation—so plainly manifesting the republican principle of ant-life, and proving that each individual enjoys a far greater freedom of thought, than, for example, a bee, strictly bound down to one method—had already been made by Huber, who resumes the account of his observations on wall-building ants, and especially on *F. fusca*, with the following words, applicable to the building of all ants without distinction:—

"I am convinced by this and a thousand similar observations, that each ant builds for itself independently of its companions. The first which conceives an easily executed plan at once lays the foundations. The others have only to continue what the first began, and learn by looking at the commenced work. But all understand how to improve their work according to circumstances; they have no other cutting instruments than their teeth, no other guide than their feelers, and no other trowels than their forefeet, of which last they make use in really marvellous fashion to separate the damp earth and to make it firm. Herein consist their mechanical and material appliances! They must therefore, if they follow a merely mechanical instinct, pursue a geometrical and unchangeable plan with slavish exactitude, build the same walls, and make their arched roofs of the same size; we should then be neither astonished nor surprised at their industry. But to build this irregular dome, standing on so many stories, to divide conveniently the therein contained rooms, to choose the best season for each task, especially to direct their work according to temporary circumstances, to utilise occasional advantages, and to judge between the suitability of this or of that operation, for all this capacities are required which approach real intelligence, and which prove that this creature does not act like an automaton, but like a being which is able to grasp the object of its work."

In what concerns the various methods of building among ants, Forel distinguishes not less than six or seven different species, accordingly as the nests are dug or built, or are made under flat stones or in wood and tree-trunks, under bark, or in walls and cliffs, or in houses, etc. Among these different species there may be recognised a great many diversities and transitions, as well as resemblances of building. The most frequent architectural style is that in damp earth, already described, which is worked with jaws and feet. When the work has to be done in dry earth or sand, they scrape this up with their forefeet, and throw it between their raised and spread-out hindlegs, in the same way as dogs do in scratching. If the forelegs or their extremities are cut away, they can, as we have already said,.

no longer dig. They do indeed make attempts, but succeed very badly. They then soil themselves and their larvæ, without being able to clean them, and at last give up wearied, and remain only on the top of the nest. A great mistake would, however, be made if it were thought that the scraping, digging, building, and wall-making of these little creatures is, as the philosopher's instinct will have it, the result of a tendency, changeless and compelling, placed in the little mind of the ant. If this were the case it would be impossible for some species of ants to seize as they do the nests of others and establish themselves therein, although, as Forel notices, the arrangements are often unsuitable for them. They usually change the architecture a little, especially outside, but as a rule the real builders can be very quickly recognised if the nest be destroyed. An ant-dwelling often changes its inhabitants in this way two, three, four or more times, sometimes through voluntary and sometimes through forceful desertion. We shall see later that there are also some species which do not build at all, but leave building to their slaves.

Still more against the instinct theory are the already mentioned differences of building of the same species under changes of circumstances. "Many nests or many parts of nests," says Forel, "are only provisional, while others are intended to last for years; many show very different styles in their different parts. The plan of building also is very much changed accordingly as the nest is meant for a larger or a smaller community. The outside appearance of the nest is also very different, accordingly as it is entirely closed with the exception of one hidden entrance, or opens with many holes on the outside (and this is true not only of different species but also of the same), or the population is large or small. The large thickly populated nests of *F. fusca* are found opening on all sides, and the small nests of *F. sanguinea* completely closed, while the contrary is the rule."

When danger threatens, for instance near a street, road, or court, or when the surface of the earth is very hard, the nests, mined or dug in the ground, betray their presence by no sign, and can only be discovered by chance. The dug-out earth is carried a long way off, and two or three concealed holes suffice for inlet and outlet. Many of these same ants at other places make a number of holes surrounded by the thrown up earth, the so-called "craters," and thereby keep up communication with the outer world. The earth between these craters is often so heaped up that the whole has the appearance of a walled dome, although nothing of the sort is intended. The ants which generally build walled nests know how to make mined or dug nests according to circumstances, and the reverse is also the case. Timid species (such as *Myrmecina Latreillei*) make the passages in their dwellings so narrow that two can scarcely pass each other, so as to protect themselves against the incursions of hostile ants, while *F. fusca* seeks to guard itself from the marauding excursions of slave-making species by placing the outlet from its nest as far off as possible, and uniting it to the nest by a long and very winding path. It is just the same with these nests as with the human robber-caverns or knightly castles of the middle ages, which had just such hidden ways of escape. Within the nest two different sorts of passages are often found, whereof the larger and more regular is for the use of all sexes, while the slighter and narrow one, which is often hard to find, is for the workers only.

In nests built under flat stones it is above all important to prevent the stone, after it is undermined, from sinking and injuring the nest. In order to attain this object, the ants make thick pillars and walls of earth between the rooms and galleries, and the heavier the stone the thicker its supports. For the rest, they choose their stone carefully, which must be neither too large nor too small, neither too thick nor too thin—more especially the last, so that it may not heat nor cool too rapidly. The nest is partly dug out, partly built. Of these are generally the so-called "double nests," which consist of nests built and inhabited by two or three quite different species. These live beside each other peacefully, because their dwellings are separated from each other by strong partition walls, which pass through the whole thickness of the nest from above to below. If the stone be lifted, the enmity or dislike breaks out, and nothing is more comic than to see the haste wherewith each kind seeks to save its own young in the lower rooms.

The most elegant dwellings, although constructed essentially in the same way as those already described, are those made by many species in wood, especially in old tree trunks. They take care, above all things, that the wood fibres shall remain in their natural position, but otherwise work in the most diverse fashions according to circumstances. Walls and pillars and buttresses run always parallel with the wood fibres, while a covering wall is always left outside of at least a centimetre in thickness, which is only pierced by a few entrances and outlets. If a larger hole chances to be made in this outer covering the inhabitants retire inwards, and try to stop up the opened galleries and rooms with sawdust and other materials. They also very often use in their building the spaces already made by other ants and insects, especially by the larvæ of the woodlouse.

"Imagine," says Huber, "the inside of a tree completely gnawed out, with countless more or less horizontal stories, five or six lines apart from each other, card-thick roofs carried now on perpendicular partition-walls, now on a number of little pillars, and in this way countless single rooms or cells—all in black-colored and smoked looking wood, and then you will have a picture of these ant-towns."

The Rev. H. C. McCook has published some very interesting observations on a kind of wood-working ants in America (*Camponotus* or *Formica Pennsylvanicus*, the Pennsylvanian Carpenter, "Transactions of the American Entomol. Soc., December, 1876). He describes a regular system of galleries, passages, rooms, halls, and vaults, which this ant made in a corner beam in an ironworks, in Blair Co., Pa., and says that such nests in trees had been found to attain the length of six feet. Forel has observed the greatest diversities of building among the *F. truncicola*. It can build walls like the blood-red ants, and use long beams like the turf-ants. It can erect domes and secondary domes, but can also build a nest under a stone. It can even establish itself in old tree trunks, and utilise the concentric layers of wood to make spacious rooms. The wood-building ants also know how to utilise their materials in the most diverse ways, and every material is good enough for them if it serves their purpose. If beams fail they use round poles, but these leave holes in the building and are less firm The journals of the Scientific Society of Bern, for the year 1874 (p. 41), contain the account of an incident in which the wood ants *(F. fuliginosa)*, in lack of wood, made a heap of horse-chestnut leaves into a wood-like construction in a beehive, with stories, rooms, and passages of communication. Indeed this species can not only hollow out wood, but can build walls, rooms, etc., out of sawdust, with the help of a glandular secretion. It is easily recognised by its shining black color, large head curved behind, and red-yellow feet. As Blanchard says, its buildings defy description.

The species which build their nests in rocks and walls have the most conveniences, since they can turn natural clefts and crannies to account, and without much trouble finish them with earth and sand. The making of passages between the several parts of the nest demands the greatest labor. J. T. Moggridge ("Harvesting Ants and Trapdoor Spiders," London, 1873-4, pp. 43, 44,) relates an ingenious observation very illustrative of the way in which ants arrange their buildings according to circumstances, and of their highly-developed intelligence and quickness: "The second captive colony, taken on December 28th, with the wingless green ant and quantities of larvæ, formed a strong contrast with the previous one. Here the ants at once set to work upon the construction of galleries and safety places for the larvæ below the even surface of garden mould on which I had placed them within the jar. . . . . This was done at 3.30 p.m., and by 9 that evening I found the ants most busily at work, having in less than six hours excavated eight deep orifices leading to galleries below, and surrounded these orifices by crater-like heaps, made of the earth pellets which they had thrown out. . . . . On the following morning the openings were ten in number, and the greatly increased heaps of excavated earth showed that they must probably have been at work all night. The amount of work done in this short time was truly surprising, for it must be remembered that, eighteen hours before, the earth presented a perfectly level surface, and the larvæ and ants, now housed below, found themselves prisoners in a strange place, bounded by glass and walls, and with no exit possible. It seems to me that the ants displayed extraordinary intelligence in having thus at a moment's notice devised a plan by which the superabundant number of workers could be employed at one time without coming in one another's way. The soil contained in the jar was of course less than a tenth part of that comprised within the limits of an ordinary nest, while the number of workers was probably more than a third of the total number belonging to the colony. If, therefore, but one or two entrances had been pierced in the soil, the workers would have been for ever running against one another, and a great number could never have got below to help in the all-important task of preparing passages and chambers for the accommodation of the larvæ. These numerous and funnel-shaped entrances admitted of the simultaneous descent and ascent of large numbers of ants, and the work progressed with proportionate rapidity. After a few days only three entrances, and eventually only one remained open. Yet for weeks this active work went on, and the ants brought up such quantities of earth from below that it became difficult to prevent them from choking up the bottle containing their water, which they repeatedly buried up to the neck. On January 10th the surface of the earth was raised from an inch and a-half at its lowest to three inches at its highest point above its original level, and this bulk of excavated earth represented the amount of space contained in their galleries and chambers constructed below. It was not, however, until nineteen days after their capture that the ants began to form systematic trains to carry down the seeds which I placed for them on the surface, and I suppose that they had required this time for the construction and consolidation of the granary chambers."

Moggridge also observed on this occasion the fashion in which the ants took away rootlets which pierced through their galleries, belonging to seedlings growing on the surface of the nest. Two ants worked together at this task, one of which pulled at the free ends of the root, while the other gnawed at the fibres where the strain was greatest, until the root broke. There are many other well-known instances of similar acts, mechanical and yet betraying a high degree of reflective power or calculative ability on the part of one or more ants, working together for the performance of a difficult task. Bingley relates ("Animal Biography," 6th ed., London, 1824, iv., p. 173,) how a "gentleman from Cambridge one day remarked an ant dragging along what, with respect to its strength, might be denominated a piece of timber. Others were severally employed, each in its own way. Presently this little creature came to an ascent, where the weight of the wood seemed for a while to overpower him: he did not remain long perplexed with it; for three or four others, observing his dilemma, came behind and pushed it up. As soon, however, as he had got it on level ground, they left it to his care and went to their own work. The piece he was drawing happened to be considerably thicker at one end than the other. This soon threw the poor fellow into a fresh difficulty: he unluckily dragged it between two bits of wood. After several fruitless efforts, finding it would not go through, he adopted the only mode that even a reasoning being, in similar circumstances, could have taken; he came behind it, pulled it back again, and turned it on its edge; when, running again to the other end, it passed through without difficulty."

Two exactly similar cases have been briefly told to the author. Herr F. Moll, of Worms, watched a wood-ant, which was carrying a beech-nut obliquely between its jaws, and wanted to pass through the rather narrow crevice in a gnarled root. As it did not succeed after several attempts, it retreated a few steps, laid down the husk on the ground, seized it by its narrow end and drew it easily through the crevice! Could a man have done it better? Dr. Ludwig Nagel, of Schmölle, sends the author the following: he was once on a summer day passing through a beautiful flowery valley, and on the north side of a mountain wall with a southern aspect, he came across an old decayed oak trunk; round this a swarm of ants were busied carrying building materials to the hill, while others, relieved of their loads, were coming back. He noticed one ant, laden with a bit of wood, which could not manage its task, and always fell back again after it had struggled forward a little way. A second ant perceived this, and, coming to the aid of its friend, caught hold of the wood, which was now carried to the building between the two. Rather more complicated, but therefore the more interesting, is an instance observed by Herr Albert Peifer, a merchant of Liegnitz, and related in the following words to the author:—

"A fine autumn afternoon of the year 1866 enticed me into a walk in an old fir wood; some friends were with me. An ant-hill, inhabited by the small black wood-ant, attracted my attention to the greatest degree, as indeed would have been the case with anyone who had for a long time studied these animals, and had found system and order in the apparent confusion. Amid the stirring work and movement of the little creatures I soon noticed that six or eight ants were busied in pushing away a green caterpillar, at least two inches long, which was clinging painfully to a dry twig. This appeared very difficult if not impossible for them, as the caterpillar, which was far superior in strength to the small creatures, would not yield to them and only held on the more firmly. I called back my companions, who had walked on, and willingly as we would have delivered the poor caterpillar out of the hands of its assailants, the spectacle was too interesting to spoil. As the ants now appeared to perceive the uselessness of their former efforts, they suddenly, as though directed by a single will, united their strength and forced the caterpillar on to its back, in which position it could no longer hold fast, and must await its fate, for good or for evil. Some of the ants pulled at their victim, others helped by pushing, and so it went stormily to one of the small entrances of the dwelling and down into Hades, whereinto we saw the caterpillar slowly disappear for evermore."

E. Menault ("L'Intelligence des Animaux," Paris, 1872, p. 6), when he was one day walking with Herr H. Delafoy, a scientific friend, saw a number of ants busy in drawing the wing of a cockchafer through the entrance of their nest. Various methods, tried in various ways, failed, owing to the narrowness of the opening. So the industrious little animals began to widen the opening over and over again until the passage was large enough. Another time Menault noticed a single ant pulling a rose-chafer's leg with great effort and loss of time over very unfavorable ground. At last, when she got near the nest, her sisters came to her assistance, and the booty was soon secured by their joint efforts. The way in which ants close and watch the entrances and outlets of their nests is also very interesting. Generally a number of small openings are made on the top of the ant-hill, which are closed by the workers in the evening, or when it rains, or on the approach of danger, and are re-opened when wanted. At other times the openings are made at the side and are very much hidden. Often there is only a chief opening with several secondary ones. In still other cases, no opening is visible, and is only to be found, as already mentioned, at a great distance from the nest, with which it is connected by a subterranean tunnel.

"If the nest of the yellow-red ants is watched at different times during the day," says Blanchard, "the observer will be surprised to see the alterations continually in progress. Coming in the early morning, all in the dwelling seem asleep. No entrances are to be found. At most, a few interspaces suggest the idea that the inhabitants might make their way through these little openings. Soon a few appear on the top of the dome and run about. Gradually more and more appear, and they may now be seen carrying little bits of wood with which they clear out the entrances. If the weather be fine, the most roomy passages and those which communicate with the chief chambers within the nest are opened, and now the whole community is busy and at work. When evening approaches, the industrious insects close the entrances. They wish to pass the night peacefully and sheltered from foreign attacks. The doors are quickly shut if a sudden shower penetrates. All hasten to this task so quickly that the object is attained in a few moments." These remarkable habits are very easy to observe, and yet Peter Huber was the first to describe them and to show that they were intentional. Huber also gives all particulars, relating how the ants use little beams to close or to roof their galleries, and cover the openings with them; and how they then, as the work progresses, bring ever smaller pieces to the heap. At last he exclaims: "Is not this the work of our carpenters in *petto*, when they build the gable of a house. Nature always appears to have made first the discoveries on which we pride ourselves." Huber is right. If clear observers had existed among primitive men, the knowledge which civilised nations have taken centuries to obtain would have been won far more rapidly.

Forel says that all species of ants, without exception, keep their dwellings more or less firmly shut, and only open them when and for so long as they are required for the workers, or for the swarming of the males and fertile females. For the closure they use any materials which come handy, earth, little stones, leaves, scraps of wood, pebbles, etc. Perty ("On the Intellectual Life of Animals," 1876, p. 336, 2nd ed.) relates after Hennings, that an English observer noticed that ants put a thin piece of slate over the chief entrance of their nest when rain threatened; about fifty of them were always required for pushing this backwards and forwards (?) If this observation be correct, it proves that the ants under certain circumstances know how to perform the task of closing and opening their door in suitable fashion.

The doors are often guarded by special sentries, which fulfil their important duty in various ways. Forel saw a nest of the *Colobopsis truncata*, the two or three very small round openings of which were watched by soldiers, arranged so that their thick cylindrical heads stopped them up, just as a cork stops up the mouth of a bottle. The same observer saw the *Myrmecina Latreillei* defend themselves against the invasions of the slave-making *Strongylognathus*, by placing a worker at each of the little openings of the nest, which quite stops up the opening either with its head or abdomen. The *Camponotus* species also defend their nests by stretching their heads in front of the openings, drawing back the antennæ. Each approaching enemy thus receives a sharp blow or bite delivered with the whole weight of the body. McCook noticed in the nests of the soon to be described Pennsylvanian mound-building ants, the employment of special sentries, which lay watching within the nest entrances, and sprang out at the first sight of danger to attack the enemy; and it was wonderful to see with what swiftness the news of such an alarm spread through the nest, and how the inhabitants came out *en masse* to meet the enemy. The *Lasius* species defend their large, strong, and very extensive nests against hostile attack or sieges with equal courage and skill, while other timid species seek to fly as speedily as possible with their larvæ, pupæ, and fruitful queens. There is, as Forel tells us, a regular barricade fight. Passage after passage is stopped and defended to the uttermost, so that the assailants can only advance step by step. Unless the latter are in an enormous majority, the struggle may last a very long time with these tactics. During this time, other workers are busy preparing subterranean passages backwards for eventual flight. Generally such passages are already made, and during a fight a new dome of the *Lasius* may be seen rising at a distance, it not being difficult for them to make this with the help of their extended subterranean passages and communications.

In addition to these enforced changes of dwelling, there

are also voluntary changes made tolerably frequently, partly for still unknown reasons; but these changes do not take place without previous mutual consultation and understanding. Lespès has the following on such an occurrence:—

"Ants sometimes change their dwellings because they are too much in the shade, or are too damp, or from some other unknown reason. One ant may then be seen approaching another, and holding a consultation with it, conferring by means of continual light touches on its head with its feelers. The latter then places its feet together, and awaits events. Its sister then seizes it in its jaws, and carries it to the place where it proposes to form the new building. After some time the twain return, and carry other comrades in the same way, until at last the larvæ and pupæ are picked up and transported to the new place." Some species, continues Lespès, appear to possess a richer language; for they are able to impart to each other the project of a change of dwelling, without, as generally happens, carrying their comrades to the new abode. For the rest, a fuller account of their language and of the very decided ability of the ants to communicate with each other will be given later.

Complicated and diverse as are the styles of building of European ants, they yet seem to be far surpassed herein by their sisters in tropical lands, which are far more numerous and larger both as to species and individuals, although, unfortunately, we have little exact and trustworthy knowledge of them. Many South American plains are, according to Lund, rendered quite uneven by countless ant-hills, which often have a circumference of from thirty to forty feet above ground, and of two hundred feet below. Stockes found ant-hills in North-West Australia of pyramidal shape, of a height of thirteen feet, which were so firm and broad that a man could stand on them without breaking through. The buildings of *Myrmica Texana* (Texas) are, according to Buckley, a hundred feet long, and single large rooms stretch underground from ten to eighteen feet; the earth thrown out from within looks like a regular crater. Far from the nest proper, the outlets open from long subterranean tunnels, through which the workers bring grain, leaves, and fruits to their underground city. The *Isan* in Paraguay, apparently one of the *Atta*, builds nests of clay, according to Rengger, which are twenty feet in diameter, and many feet high, and which spread far down into the ground and have many doors. From these doors go as many neatly levelled tunnels, often yards long, by which the swarms of ants march in, laden with building materials and booty.

Bingley (p. 180) relates that, in Captain Cook's expedition in New South Wales, they saw ants " as green as a leaf, which live upon trees and build their nests of various sizes, between that of a man's head and his fist. These nests are of a very curious structure: they are formed by bending down several of the leaves, each of which is as broad as a man's hand, and glueing the points of them together so as to form a purse. The viscous matter used for this purpose is an animal juice. . . . . Their method of bending down leaves we had no opportunity to observe ; but we saw thousands uniting all their strength to hold them in this position, while other busy multitudes were employed within, in applying this gluten, that was to prevent their returning back. To satisfy ourselves that the leaves were bent and held down by the efforts of these diminutive artificers, we disturbed them in their work ; and as soon as they were driven from their station, the leaves on which they were employed sprang up with a force much greater than we could have thought them able to conquer by any combination of their strength." The observers were punished for their destructiveness by being severely stung by the ants. A similar account is given by the same observers of two other species, of which one made their nests in branches, like the European wood-ants, the other by hollowing out the roots of living trees.

CHAPTER VII.

Road-Making.

THE ants show almost more cleverness, circumspection, and acuteness in utilising given circumstances or in conquering natural difficulties than even in building their dwellings, in making roads—which in their various businesses carried on in and out of the home are naturally of the greatest importance to them. The subterranean tunnels serve sometimes as means of communication with other ant colonies, or with the several parts of an extensive nest; sometimes as has been said, to hide the entrance of the nest, sometimes as roads to a place where food is to be obtained, as for instance sweet sap or the milch-cow Aphides: for all these purposes open uncovered ways are also used, just according to circumstances. They always choose the shortest way and the most convenient and advantageous method of obtaining their object. They will also use an already prepared ground, on which they can pass backwards and forwards undisturbed for a certain distance, without actually making a road, as for example the base of a wall, or a fence, or the edge of a walk. Where this cannot be done, a special path is made, levelled and cleared from all obstacles, such as dried leaves. Thus regular high roads are built in wooded meadows from one nest to another; the grass is first mown, and then a firm foundation of cement and sand is laid, and on this again a raised embankment, on which they run actively to and fro. In a wood, where the making of the road itself is very easy, but where falling leaves and such like obstacles block the way, the ants give it a certain width, often of as much as two decimetres, or deepen it a little, while in meadows wherein the construction is, as described, difficult but lasting, the road is only from four to six centimetres in width. They do not begin their roads, as might be expected, from their nests outwards, but spread themselves over all the lines whereon roads are to be made (as in the building of a new nest), begin the work simultaneously at all points, and then behave just as men do in making a railway, high-road, or similar work. The roads sometimes extend to a distance of eighty and even of a hundred yards from the nest, and eight or ten of these will be made from a large nest. The wood ants, as a rule, make no roads, as their passage from one tree to another is attended with no difficulty.

The covered ways, galleries, or tunnels are, as a rule, built by those ants which visit their milch-cows, the Aphides, secretly, and need to hide these from other ants or from their countless other enemies. If the tunnel passes through a quiet out-of-the-way place, where no danger is to be feared, they leave off the roof and the road is open. Forel saw such a road, two centimetres broad and one centimetre high, built of earth and covered in, which was taken up a low wall and down again on the other side, merely to secure a safe way from a court into a garden. Sometimes they walled round the stalks or stems of the plants on which they kept Aphides, so that the latter were quite shut in, and made rooms for them of the leaves of the plants. *Lasius brunneus*, the brown ant, almost lives by keeping very large cochineal insects, and protects its roofs by help of their rotted skins. In the same way Kermes, and especially the gall-insects, are sometimes regularly walled in by ants. Their prison is generally tolerably large, and has small openings, by means of which the caretakers slip in and out, while the outgoing of the prisoners is prevented. Forel saw such a prison, shaped like a cocoon, about a centimetre long, hanging on an oak branch, and containing Aphides, which were carefully looked after by the ants. In the chapter on " Cattle-care and milking" more will be said on this.

Forel had special opportunities of observing that ants are able to perfect themselves in their tunnel-making (as in their various other works). He kept a colony of *M. cæspitum* (turf-ants) imprisoned in a tray walled round with pulverised gypsum, so that the gypsum gave way under climbing and the ants fell down. This went on for a fortnight, until it struck the ants to try to escape by making a tunnel through the powder. Many efforts were rendered futile by the fall of the loose gypsum; but at last, after many failures, the little creatures managed to make a way right through the thickness of the wall. Forel then blocked the tunnel by slightly shaking the gypsum. But the ants had clearly become perfect at their work, for they quickly remade the tunnel through the mass that had fallen in. Forel then left them at peace, and they escaped with their larvæ and pupæ. Perhaps yet more interesting than the roads, are the stations which the ants make for the shelter of the provisions, or of the workers when the way is long. They are generally small nests, for the reception of the workers when wearied or overpowered by the sun, or for a warm shelter if they are benighted. If a sudden shower of rain overtakes them, they also resort to them. Sometimes these are mere holes in the earth, but at other times they are thorough little nests, dome-covered, which gradually can be made to serve as real nests. Mention will be made later of the way in which harvesting ants make regular depôts for grain on the road.

In addition to these stations, there are also suburbs and dependencies of the large nests, which receive the surplus population. New nests thus often spring up under such circumstances in the neighborhood of the old one, and this enlargement gradually grows to an enormous circumference round the original spot. Forel saw a colony of *F. exsecta*, in a wood-clearing on Mont Tendre, which consisted of more than two hundred nests, and covered a space of 150—200 square metres. A similar, but somewhat smaller colony, of *F. pressilabris*, is found in the Petite Salève near Geneva. The whole space is covered with stunted shrubs, on which the ants keep their Aphides. All the members of such a colony, even those from the furthermost nests, recognise each other and admit no stranger. This fact, together with the great number of inhabitants, which all hold together, gives such a colony unusual strength; and one can, with Huber, unhesitatingly compare their several nests to the several cities of one and the same country, or still better with those of one and the same Republic. The whole realm rises to repel an attack from without, and the subterranean as well as surface means of communication between the different nests make possible a rapid concentration of the marshalled war-forces on a single point, just as is the case in a human kingdom by means of many and well-arranged railways.

But all this falls far short of the description given by the Rev. Mr. McCook, in the Trans. of the Amer. Entomol. Soc., Nov., 1877, of the "Ant Town" he discovered in the Alleghany mountains, North America. This town, built and inhabited by the *F. exsectoïdes* (Forel), or mound-making ant of America, lies on the eastern side of the Brush mountain, Pennsylvania, not far from Hollidaysburg, and consists of no less than 1600—1700 single colonies or nests, which rise in cones to a height of from two to five feet and have a circumference of from ten to fifty-eight, covering a plain of about fifty acres in extent. Beneath each nest is an extensive system of subterranean galleries and roads, which often stretch to a length of sixty feet to the neighboring trees and food resorts. McCook reckoned that these remarkable buildings, compared with the size of their architects, are a thousand times as large as the buildings of men. The inhabitants of all these nests are bound together in strict alliance and never carry on war against each other, and may therefore be rightly regarded as a large republican commonwealth, or a kind of Federative Republic. Injuries suffered by these nests or hill-dwellings, are very rapidly repaired by their united forces. Small animals, thrown on the surface of the nests, as large spiders, snakes, etc., are torn to pieces with equal rapidity, and gnawed down to the chitin, or skeleton.

To return once more to the covered ways, or tunnels; these are built for another reason than that named above—viz., for protection against the sun. As much as ants love the mild spring or autumn, and know how to utilise it for the benefit of their young, so much do they avoid the hot midday sun of summer, which quickly shrivels up their little bodies. On very hot days they therefore work in the mornings and evenings, and take a siesta in the middle of the day. They thus behave exactly as men do in hot countries and on hot days. Lespès noticed this with *Atta barbara*, one of the species of harvesting ants of the Mediterranean shores, and Moggridge remarked the same. Lespès also saw this species working by moonlight at night, while Moggridge saw them working "when there was neither moon nor stars to be seen." This, as well as the observations of Gould, Huber, Kirby, Ratzeburg, Forel, and others, resolves the question so much ventilated since

Aristotle, as to the night work of ants. Aristotle held the view, so much challenged later, that ants worked at night during the time of full moon, and this has since been proved to be thoroughly correct. Forel watched a dwelling of the turf-ants, which scarcely left their nest during the whole day through a very hot period in the month of July, and came out in thousands during the evening, swarming over all the roads and visiting the Aphides in the trees. He also found them on a very dark night by means of a lantern, in company with their beloved nurslings on trees and bushes. This activity continued during the whole night and had not decreased by the early morning. Forel repeated this observation with several other species, as well as in his vivarium. When it was very warm, he saw its inhabitants busy the whole night through; they slept on the contrary in cold weather. In the spring, things are just the opposite; the ants then, as a rule, leave their nest about eight or nine o'clock in the morning, and return at about five or six. The ants which work in the dark appear to find their way by help of their antennæ, while such as have comparatively smaller feelers and more highly-developed eyes (as *Polyergus rufescens*, *F. rufa*, etc.) generally prefer day to night. Their movements at night are also slow and measured, while they move rapidly and hastily by day. Mr. McCook noticed his ants at work equally day and night; frost and cold caused them to fall into a kind of doze, and all work and action stopped. He did not see them come out at all during the winter; they then remained in the deepest and warmest part of the nest. There is, after all, nothing unusual in the night work of ants, seeing that all that is done within their dwellings goes on in complete darkness. The wall-building ants especially resort to night labor, because their walls do not then dry so rapidly, and thus attain greater firmness; while the species which do not build prefer the day. In rainy weather or on damp cloudy days, the builders also may be seen at work. But if the little architects are suddenly overtaken by the hot sun during their expeditions, or if their road from being in the shade leads over a place upon which the sun is shining brightly, they then quickly roof it over, or roof the exposed part, with a gallery made of earth and saliva; or, if circumstances lend themselves to the proceeding, they tunnel through the ground, and pass backwards and forwards under its shelter. This method of road making is specially used by the driver or marching ants of West Africa (*Annomma arcens*), which are very sensitive to the burning rays of the African sun, and therefore generally work during night or under a cloudy sky. They have no fixed abode, and do no regular building, but seek shelter under hollow tree-roots, overhanging rocks, and other shady places. They march quite regularly through the African woods in large closed columns two inches wide, and seize all living things that they find on their way. They are therefore a much dreaded plague. The ants laden with food or larvæ keep in the midst of the ranks, while outside or on either flank march the soldiers, or officers—that is, individuals with enormously large heads and powerful jaws, which carry no burdens but only watch over the order and safety of the column, act as scouts, bring in runaways or laggards, and defy any attack. Very rarely, however, does either man or beast venture upon one.

If it chances to one of these processions to be delayed in the open until late in the morning, either by rich booty or by other hindrance, the travellers quickly cover their path with an arched roof of earth or dirt, moulded together with saliva. If long grass, on the other hand, yields them sufficient shelter against the destructive sunrays they cease this work. These ants live by making marauding excursions, and they take their name from the fact that they drive all living things before them. Even large animals suffer from their attacks, and it is said by the natives that the boa-constrictors, when they have crushed their victim in their deadly coils, before they begin to devour it search over a wide circle of at least a quarter of a mile in diameter, in order to see if an army of driver ants is on the march: if one is near they glide away and leave their prey to the ants. These suck the liquid juices from slain animals and then pull the flesh, piece by piece, into their hiding place. If, during their nightly marches they penetrate into a human abode, it only remains for the dwellers to take to their heels and fly at once, and this is done the more willingly as all the vermin previously settled in the house, such as rats, mice, snakes, lizards, cockroaches, spiders, bugs, etc., are totally destroyed by the ants. It is true that pigs, fowls, etc., that have been left behind by forgetfulness or lack of time also fall victims to them. When the ants have departed, the owners come home again. If their hiding place is flooded by a tropical rainstorm, the ants crowd closely together into a large heap, take the younger and weaker in the middle, and pass over the water till they reach a dry place. Over smaller streams they form a living bridge when on the march, by clasping each other firmly and so forming a chain by which the host crosses. By help of such chains also they let themselves down from trees.

In South America there is a travelling, marching, or foraging kind of ant, with exactly similar habits, about which the English naturalist, Bates, in his "Travels on the Amazon," has published a complete account, and of which more will be said further on.* It is, however, a suitable place for dealing more fully with another non-European species, which has a special interest. It is the Brazilian travelling or visiting ant, *Myrmica*, or *Atta*, or *Æcodoma cephalotes*, sometimes called *Sa-uba*. They are usually known by the name of umbrella-ants, because they are seen marching in long crowded columns, each holding up between its jaws a generally round bit of leaf, about the size of a sixpence. They, however, do not use these bits of leaf as sunshades, as was thought, but for roofing in and covering their very extensive domed dwellings. The *Sa-uba* is a terrible plague, for it strips the most valuable trees, namely the coffee-plants and orange trees, of their leaves, and by its countless numbers makes agriculture almost impossible in some places. Lund relates that he one day passed a tree in full leaf, and heard a patter like heavy rain although the sky was quite clear. He went closer, and saw an ant hard at work on every leaf-stalk. When the leaf-stalk was gnawed through the leaf fell to the ground. A still more remarkable scene was going on at the foot of the tree. The ground was covered with ants, which seized the leaves as they fell in heaps, and bit them in pieces. In the space of an hour all was over. The tree was stripped, the leaves cut up, the pieces carried away. The famous traveller, Dampier (see " Bingley," p. 179), saw similar trains of these umbrella-ants, which, as he put it, looked like a stream of moving pieces of leaves, under which the insects

---

* [The migrating ants of Cayenne, the "Ants of Visitation" (Mme. Merian), and the Chasseur-ants of Trinidad, have similar habits. TR.]

carrying them are quite hidden. Yet they moved very rapidly, and the whole road or path looked perfectly green.

Dr. Fr. Ellendorf, of Wiedenbrück, who has lived many years in Central America, writes as follows to the author on this remarkable animal : " The umbrella-ants belong to the most interesting of the ant-species. I often met them in Costa Rica at the beginning of the dry season, hurrying in millions towards the coffee plantations, or returning to the domestic hearth, each carrying a little green flag on its head. I often watched their busy, industrious ways, and once let myself be so led away by them that I tied up my mule and walked along the procession to find the nest. But as at the end of half-an-hour I had not succeeded, I turned back for want of time. A few years ago, later, I had more time and opportunity for watching them. Soon after the commencement of the dry season the grass on the hill-sides is dried up by the rays of the tropical sun, and the ants then begin to visit the coffee plantations. What work and patience to provide for their wants! It would be quite impossible for them to creep even through short grass with loads on their heads for miles. They therefore bite off the grass close to the ground for a breadth of about five inches, and throw it on one side. Thus a road is constructed, which is finally made quite smooth and even by the continual passing to and fro of millions upon millions night and day. As soon as they have reached the plantation they climb up the trees, and each ant in about a quarter of an hour has cut out of a leaf by its mandibles a crescent shaped piece, half an inch long ; it holds this firmly over its head and sets off homewards. If the road is looked down upon from a height on the way back, with these millions thickly pressed together and all moving along with these green bannerets on their heads it looks as though a giant green snake were gliding slowly along the ground, and this picture with its greenish-yellow background is so much the more striking that all these bannerets are swaying backwards and forwards.

"At the end of the war in Nicaragua I lived for a long time in the little town of Nivas. On an excursion after butterflies I one day met a laden train and seized the opportunity of watching the creatures more closely. I first wished to see how they would manage if I put an obstacle in their way. Thick, high grass stood on either side of their narrow road, so that they could not pass through it with the load on their heads. I placed a dry branch, nearly a foot in diameter, obliquely across their path, and pressed it down so tightly on the ground that they could not creep underneath. The first comers crawled beneath the branch as far as they could, and then tried to climb over, but failed owing to the weight on their heads. Meanwhile the unloaded ants from the other side came on, and when these succeeded in climbing over the bough there was such a crush that the unladen ants had to clamber over the laden, and the result was a terrible muddle. I now walked along the train, and found that all the ants with their bannerets on their heads were standing still, thickly pressed together, awaiting the word of command from the front. When I turned back to the obstacle, I saw with astonishment that the loads had been laid aside by more than a foot's length of the column, one imitating the other. And now work began on both sides of the branch, and in about half an hour a tunnel was made beneath it. Each ant then took up its burden again, and the march was resumed in the most perfect order. The road led towards a cocoa plantation, and here I soon discovered the building which I afterwards visited daily. As I again went thither one day I was met, at a considerable distance from the nest, by a closely pressed column coming thence, and all the ants laden with leaves, beetles, pupæ, butterflies, etc.; the nearer I came to the nest, the greater was the activity. It was soon plain to me that the ants were in the act of leaving their dwelling, and I walked along the train to discover the new abode. They had gone for some distance along the old road, and had then made a new one through the grass to a cooler place, lying rather higher. The grass on the new road was all bitten off close to the ground, and thousands were busy carrying the path on to the new building. At the new home itself was an unusual stir of life. There were all sorts of laborers—architects, builders, carpenters, sappers, helpers. A number were busy digging a hole in the ground, and they carried out little pellets of earth and laid them together on end to make a wall. Others drew along little twigs, straws, and grass-stalks, and put them near the place of building. I was anxious to know why they had quitted their old home, and when the departure was complete, I dug it up with a spade.

At a depth of about a foot and a half I found several tunnels of a large marmot species, the terror of cocoa planters, because in making their passages they gnaw off the thickest roots of the cocoa plants. The interior of the ant hill had apparently fallen in through these mines. Unfortunately I was unable to follow further the progress of the new building, for I was obliged to leave the next day for San Juan del Sur. When I returned at the end of a week the building was finished, and the whole colony was again busy with the leaves of the coffee plants."

The most detailed accounts of these remarkable ants are given by H. D. Bates ("The Naturalist on the River Amazon," London, 1863). According to him, also, the *Sa-uba* is a terrible pest in Brazil, and makes agriculture impossible in many districts. Their workers fall into three distinct classes, and vary in size from two to seven lines. The domes of their nests, often forty yards in circumference, rise only two feet from the ground, and guard the entrances to their extensive underground galleries. These entrances are generally closed, and are only opened under special circumstances. A large hollowed out chamber, of from four to five inches in diameter, is found in the interior of the nest. The habit of the *Sa-uba* of cutting off and carrying away large quantities of leaves has long been known. The columns then resemble a large mass of moving leaves. Near their nests are found heaps of such rounded bits of leaves, and these are always carried off on the following day. Bates often saw them at work, and noticed that they cut a half-circle in the upper face of the leaf with their razor-like mandibles, and then pulled it away from the uncut side by a sudden jerk. It was often thrown down to the comrades waiting below, but as a rule each ant marched off with its own piece. Cultivated trees attract them more than any other. The leaves serve as roofs for their dwellings and as a protection against tropical rains for their young. Some of the laborers pull the leaves along; others put them in their right places and cover them with a layer of earthy matter brought up from the depths of the nest.

The Rev. H. McCook (*loc. cit.*, 1879, p. 84) watched the same kind of ants in Texas, where they are called Brazilian or cutting-ants. Their trains seemed to him like a procession of Lilliputian Sunday-school children, carrying their flags over their heads. The door of the nest was opened for each excursion, and then closed again by the help of dry twigs, leaves, and other rubbish. The chief work was done by the larger, the lighter by the smaller individuals, or "castes." The opening and closing of the nest was performed, as a rule, mornings and evenings, at a regular hour and in regular succession. McCook also observed a systematic division of labor in the leaf-cutting and the carrying away of the pieces, the "soldiers" only acting as guards and as leaders. Trees with thin leaves were preferred by the wise little animals, and the leaves were made into a kind of paper-like mass, in the interior of the nest, by means of their jaws and their saliva, so as to serve for building the nest and for the making of single cells or larval chambers. Possibly they live on the juices of the leaves. The larvæ are cared for, and the mining or working in the earth is chiefly done, by the smaller individuals or castes, the so-called minors, while the larger ones do the outside work and are helped in the daily opening and shutting of the doors by the smallest ones. These smallest ants are, according to McCook, very brave and often do yeoman's service to the large-headed soldiers.

The subterranean workers of this remarkable genus are very clever. The Rev. H. Clark reports from Rio de Janeiro, that the *Sa-ubas* have made a regular tunnel under the bed of the river Parahyba, which is there as broad as the Thames at London, in order to reach a storehouse which is on the opposite bank. Bates tells us that close to the Magoary rice-mills, near Para, the ants bored through the dam of a large reservoir, and the water escaped before the mischief could be remedied. In the Para Botanical Gardens an enterprising French gardener did everything he could to drive the *Sa-ubas* away. He lit fires at the chief entrances of their nests, and blew sulphur vapor into their galleries by means of bellows. But how astonished was Bates when he saw the vapor come out at no less a distance than seventy yards! Such an extension have the subterranean passages of the *Sa-ubas*.

Ants which have invaded houses and settled therein are attacked in Brazil by the so-called "Ant-master" in similar fashion; he pumps smoke into their passages to kill them, after he has stopped up all the outlets. As soon as the ants mark that something is amiss, they pick up their eggs, larvæ, and pupæ and try to escape. But when they find that all the outlets have been stopped, they at once change their plan, lay down their burdens, and begin to dig new ways out. These also are stopped, and they try again and again until all are killed.

McCook saw near Austin (Texas) a hole of the *Sa-uba* twelve feet in diameter and fifteen feet deep, from which more than a hundred subterranean passages, hundreds of feet in length and from two to six feet deep, ran to the nearest trees.

In addition to the destruction wrought by the *Sa-uba* on the trees and in the ground, the natives suffer by the plundering at night of their corn and flour stores. Bates was at first inclined to lend no credence to current tales of this nature, but he was soon convinced by his own suffering therefrom. While he was living in an Indian town on the Tapajoz, his servant awaked him many hours before sunrise with the cry that the rats were plundering his baskets filled with farinha or mandioca-meal.* As the article was then rare and very dear he got up, and found that the noise arising from the robbery bore very little resemblance to that made by rats. He struck a light, and saw a broad column of thousands of *Sa-uba* ants, which were carrying the contents of his baskets out through the doors as quickly as possible. Each one carried a grain of farinha, which was often larger and heavier than its bearer, and it was amusing to see the smaller ones regularly tumbling over with their load. The baskets were quite covered with them, and hundreds were busy cutting out the dried leaves which served as lining to the baskets. It was this that made the noise that had been heard. The servant declared that they would quite empty the two baskets during the night, if they were not driven away. They therefore tried to kill them with wooden clogs, but new hosts took the places of the slain. At last Bates drove them away with gunpowder, which he scattered on the ground and fired. McCook also saw the Texan cutting-ants plunder a wheat-store and carry the grains to their nest.

---

* [Farinha, or mandioca meal, consists of grains resembling tapioca, and is obtained from the same root. Tapioca is pure starch, while farinha is starch mixed with woody fibres. TR.]

CHAPTER VIII.

HARVESTING ANTS.

BATES could not understand, judging by what he says, what the ants wanted with the hard, dry mandioca grains. If he had been as familiar with harvesting ants as we now are—thanks to the interesting publications of his countryman, J. T. Moggridge—he would not have been left in doubt. It has been already shown in the historical review given above (chap. v.) that the habit of ants in Southern countries of gathering corn and using it as stored up food, was well known in ancient times; and the belief remained until later observers (Swammerdam; Gould, Christ, Latreille and others) rose up against it, and declared that the whole account of harvesting ants was a fable. Huber himself spoke most decidedly against it, and was, indeed, supported by very good grounds. He maintained that, to begin with, the parts of the ant's mouth were very unfitted for eating hard corn, and that they could only feed on soft matters or fluids; and that, secondly, a storing up of food for winter provision was quite unnecessary, because they hibernated during the cold weather, and needed no food. But if by chance, says Huber, a warm day comes in the winter and wakes them up, they always have at hand an adequate number of Aphides, which will also have been aroused by the warm rays of the sun, and from which they can obtain nourishment. So far as the strong mandibles of the ant are concerned, we have already said that they serve as weapons or tools, and never for eating.

These views of Huber, the distinguished observer, silenced for a long time all contradiction, and this the more because ants were never seen to store up corn in Northern lands, save that occasionally they picked up grains of corn, like other things, and took them as building material to the nest. The idea that ants were completely idle during the winter must have been very generally held in Shakspere's time, for the great bard makes the fool say to Kent, in "King Lear" (act ii., scene 4): "We'll set thee to school to an ant, to teach thee there's no laboring in the winter." The facts given by Huber are perfectly correct, but the deduction therefrom is incorrect. Huber and the earlier opponents did not remember that the ancient stories came to us from Greece and from the East, where the habits of the ants are different from those of the North, in consequence of the difference of the climate. But in the North also we are not quite without harvesting ants, and if Huber had searched more carefully he might have found the solution of the problem on the Petite Salève, near Geneva, as Forel did. Of the two chief harvesting species in Europe (*Aphænogaster*, or *Atta structor*, and *Atta barbara*) the first is found in Switzerland. This species, indeed, there collect different grains, which also serve for food after the starch-flour contained in them has been partly turned into sugar as the result of germination. But although the two species named are rare in the North, they are more numerous in Southern Europe and especially on the shores of the Mediterranean, and both Lespès and Moggridge have observed and accurately described their habits. They do not hibernate there, owing to the warmth of the climate, any more than do Northern ants if they are kept in a warm room, and therefore need the winter-stores which they keep in the interior of their nests in special granaries. Lespès often saw little heaps of refuse in front of their nests, which they had made of the husks after eating the corn, soft and swollen by germination in the interior of the warm, damp nest. At least the little radicle—the tenderest and sweetest part of the grain—was always, says Lespès, eaten off. Moggridge noticed these same rubbish heaps. In gathering corn they observe, according to Lespès, that great economical principle which plays so large a part in the industrial and general life of man, and which we have learnt to recognise as the chief reason of their successful industry in other ant labors, such as digging, building, earth and leaf carrying, nursing the young, etc.—the principle, namely, of division of labor—and they carry it so far that if the road from the place where they are gathering their harvest to the nest is very long, they make regular depôts for their provisions under large leaves, stones, or other suitable places, and let certain workers have the duty of carrying them from depôt to depôt. Lespès sometimes found two or three such depôts on a single road. This is yet more strikingly illustrated by an observation of Moggridge, quite analogous to that which we have already given from Bates on the leaf-cutting species. He saw some ants climb up the haulms of the grain-bearing ears, and shake or throw down the grains, while others waiting below took up the fallen grains and carried them towards the granaries. But they only took them to the entrance of the nest, where other comrades were in waiting, and pulled the grains inside. Moggridge also tried to deceive the little robbers by scattering little china beads of the size and color of the grain in their way. These were at first picked up and carried off. But the wise little creatures soon found out their mistake, and left the useless beads alone. But let us rather allow Moggridge to tell his interesting experiences himself, in a somewhat compressed form:—

"I had scarcely set foot on the *garrigue*, as this kind of wild ground is called to distinguish it from meadows or terraced land, before I was met with a long train of ants, forming two continuous lines, hurrying in opposite directions, the one with their mouths full, the others with their mouths empty. It was easy enough to find the nest to which these ants belonged. . . . . The nearly continuous double line measured twenty-four yards. Even this gives but an inadequate idea of the number of ants actively employed in the service of this colony, for hundreds of them were dispersed among the weeds on the terrace, and many were also employed in sorting the materials and in attending to the internal economy of the nest. . . . . It is not a little surprising to see that the ants bring in not only seeds of large size and fallen grain, but also green capsules, the torn stalks of which show that they have been freshly gathered from the plant. The manner in which they accomplish this feat is as follows. An ant ascends the stem of a fruiting plant, of Shepherd's Purse *(Capsella bursa pastoris)* let us say, and selects a well-filled but green pod about midway up the stem, those below being ready to shed their seeds at a touch. Then, seizing it in its jaws, and fixing its hindlegs firmly as a pivot, it contrives to turn round and round, and so strain the fibres of the fruit-stalk that at length they snap. It then descends the stem, patiently backing and turning upward again as often as the clumsy and disproportionate burden becomes wedged between the thickly-set stalks, and joins the line of its companions on their way to the nest. .... Two ants also sometimes combine their efforts, when one stations itself near the base of the peduncle and gnaws it at the point of greatest tension, while the other hauls upon and twists it. .... I have occasionally seen ants engaged in cutting the capsules of certain plants drop them and allow their companions below to carry them away; and this corresponds with the curious account given by Ælian [see pp. 55, 56]. .... Occasionally [in addition to seeds] one or two may be detected carrying a dead insect, or crushed landshell, the corolla of a flower, a fragment of stick, or leaf, but I have never seen Aphides brought into the nest or visited by this ant or by *Atta structor*. It sometimes happens that an ant has manifestly made a bad selection, and is told on its return that what it has brought home with so much pains is no better than rubbish, and is hustled out of the nest and forced to throw its burden away. .... I have often amused myself by strewing hemp and canary seed or oats, all of which form heavy burdens for the ants, near their nests; and it is a curious sight to see the eagerness and determination with which they will drag them away. It is interesting also to note how on the following day the husks of these seeds will appear on the rubbish heap, or sometimes, after a shower of rain, they will be brought out by the ants with the point of the little root (the radicle or fibril as the case may be) gnawed off. It frequently happens that on the wild hillside the position of a nest of *Atta barbara* is indicated by the presence of a number of plants growing on or round the kitchen midden, which are properly weeds of cultivation, and strangers to the cistus and lavender-covered banks of the garrigue. These have sprung from seeds accidentally dropped by the ants, and which they had obtained from the lemon terraces. .... The large mounds which may frequently be found at the entrances of their nests, are nothing more than the rubbish heaps and kitchen middens of each establishment. These consist in part of the earth pellets and grains of gravel which the ants bring out from their nests when forming the subterranean galleries, but principally of plant-refuse, such as the chaff of grasses, empty capsules, gnawed seed-coats, and the like, which would occupy much space if left inside the nest. While an army of workers are employed in seeking and bringing in supplies, others are busy sorting the materials thus obtained, stripping off all the useless envelopes of seed or grain, and carrying them out to throw away" (pp. 16—21). In sheltered places these heaps grow to a considerable size.

In October, 1873, Moggridge found near the entrance of a nest of *A. structor* such a rubbish heap, round in form, which was twenty-seven inches in diameter and two inches in thickness, its whole amount showing the existence of a large quantity of seed within the nest. Indeed, when he opened and more closely examined several nests, he found masses of seeds carefully stored in large chambers. The floor of these granaries is carefully levelled and cemented, and can at once be distinguished by its appearance from the surrounding ground. The rooms themselves were exceptionally large and regularly shaped, and were generally about the size of a watch-pocket. In each were found, on an average, a hundred grains, and the quantity of grain in a nest can be calculated, as it has often from eighty to a hundred such single rooms, at about a pound or more. The seeds belong to various kinds of plants, and Moggridge found in one nest he opened the seeds of twelve different species, belonging to at least seven distinct genera. The ants chiefly gather the grains of cultivated cereals, doubless because they contain more nourishment.

Moggridge was most surprised as to the still not quite explained way in which the ants prevent the seeds from germinating and growing. It is a matter of course that seeds cannot remain long under the ground, within the damp and warm nest without germinating, and sprouting forth into grasses or herbs, and this would defeat the object with which the ants garner them. But Moggridge found, among thousands of seeds, in twenty-one nests, examined for this purpose, only a few which had germinated, and of these nearly the half were so mutilated that the growth would soon cease. There can, therefore, be no doubt that the ants, by some unknown way of treating the seeds, make them incapable of germination, at least for some time—namely, for weeks or months. Moggridge was not in a position to solve the riddle by the experiments and trials he made. If he kept the ants away from certain granaries the grain therein began to sprout, proving that it was not outside circumstances but some action on the part of the ants themselves which prevented germination. In deserted or separated parts of the nest, the seeds sprouted into grasses.

Perhaps the ants manage to purposely delay the sprouting with some gummy substance, by mechanical stopping up of the so-called eye of the seed, whereby damp penetrates within. When the time comes for them to make use of the seeds as food, this substance is then removed, and the seed purposely moistened and caused to germinate. But in order to prevent the further growth which would render the seed useless for food, the ants bite or gnaw the growing sprout or radicle, and then bring out the thus changed seed to dry in the sun. They are then again granaried. When they have become wet by falling rain, they are submitted to a drying process in the same fashion.

It is well-known that seeds are much altered by germination, especially cereals, so that the starch contained in them is changed into sugar and gums. At the same time the hard integument is broken, the whole grain swells and becomes soft. It is then in the condition needed and desired by the ants, which eat the soft part, especially their favorite sugary matter, or satisfy therewith the large claims of their growing larvæ, and leave the shell or husk in the form of refuse. This refuse forms the chief part of the rubbish heaps described above. This whole process is just the same as that adopted by brewers towards malt of barley or corn, so that it cannot be doubted that ants are thoroughly acquainted with one of the most important branches of human ingenuity and industry—nay, that according to all appearances they were acquainted with it before men appeared on the surface of the globe. "Instinct" cannot have thus taught them, but only experience, and the methodical application of this occasional experience to an appropriate end can only be the result of a conscious act of reflection, which in certain genera has gradually become a heritable mental habit. They are, however, prudent enough not to allow themselves to be so completely controlled by this inherited mental habit, as not to spare themselves the troublesome harvesting and garnering of corn, when they can obtain it—or, as we now prefer to say, annex it—more conveniently by thieving or plundering, either from human stores and granaries, or from the granaries of their own relations. Moggridge saw a flourishing colony of *Atta structor* in the main street of Mentone, which had settled itself down happily at the door of a corn chandler, and carried up the grains falling from oats and wheat. Another nest, in another part of the town, obtained a part of its food from the canary seeds scattered by some caged birds above. Moggridge also discovered some secret robber-paths from several nests to human granaries lying near, and the establishment of such passages is the more possible as the species noticed by him, as Moggridge showed, are able to make tunnels and galleries in hard stone (sandstone)!

But the harvesting ants, like men, find it most convenient and handiest to practice robbery and plundering on their own brothers and relations. Perhaps they are also driven in this direction by the warlike and battle-loving spirit innate in most species of ants. The shining, coal black *Atta barbara* chiefly distinguishes itself in this way, and often carries on plundering fights that last for days and weeks. Moggridge saw such a struggle that went on from the 18th of January to the 4th of March. As often as he visited the place during this time, he saw only scenes of battle and plunder. The hostile nests lay about fifteen feet apart, and the ants fought in the bitterest fashion over every grain of corn. The ground was always covered with the slain and the terribly mutilated. Most frequently the hinder part of the body was bitten or dragged off. But the remaining portion, consisting often of only the head and a few legs, clung to the disputed grain with spasmodic force, and was often pulled away with it by the victor. Only when a fighter succeeded in injuring or tearing off the feelers of its rival, did the latter appear quite palsied and defenceless.

Moggridge soon became convinced that the dwellers in the upper nest plundered the granaries of the lower one, while the dwellers in the latter tried to get back their stolen seed, or in return to steal some food. The thieves which had commenced the strife were clearly the stronger, and whole trains of laden robbers took their way to the higher nest, while comparatively few seeds found their way, in return, to the lower. Even this only succeeded occasionally, for some of the robbers were posted in front, which received those coming down from the upper nest, and after they had conquered all the obstacles in their road, chased them away from 'their booty and took it back again. After the 4th of March Moggridge observed no further hostilities, although the lower, the plundered, nest was not deserted. But when he visited the spot again in the October of the same year, he found the latter quite empty and silent, while the robber nest appeared to be full of life, and its granaries quite filled. In another case, in a battle that had gone on for thirty-one days, the plundered nest was finally wholly deserted, and when Moggridge opened it he found the granaries totally empty.

Harvesting ants seize everything in time of need, even quite other food if they can get it; specially will they take dead insects. Moggridge saw how a dead grasshopper, which he had thrown in front of a nest, was carried inside. It was too large to pass through the door, so they tried to dismember it. Failing in this, several ants drew the wings and legs as far back as possible, while others gnawed through the muscles where the strain was greatest. They succeeded at last in thus pulling it in. Another day Moggridge saw the wings of the animal appear on the rubbish heap. Harvesting ants kept in captivity devoured dead house flies and the larvæ of bees and wasps, but never troubled themselves about the Aphides or similar insects, so much liked by other species. The sweet juices or plant excretions, so much sought after by most ants, appear also to be despised by them. On the other hand Moggridge saw them gnaw the joints of a dead lizard, and one day watched a fight between two middle-sized individuals of the genus *Atta barbara* and a thick grey caterpillar an inch long, which in vain made the most vigorous efforts to escape from its little tormentors. Moggridge took the little group home and put them in spirit. But even death could not compel the two robbers to loose their grip of their prey. Moggridge, by throwing an artificial light on ants kept in captivity, had also the opportunity of seeing how they eat the corn or its contents. In a group of ants he discovered one holding fast to a white round lump. This lump was seen to be the mealy part of a millet seed, and Moggridge saw how two or three other ants tore off little particles with their toothed mandibles, and then conveyed them to their mouths. They repeated this often, before they gave place to their comrades.

It follows from this that harvesting ants can also eat dry substances, and are so far an exception to their fellows, which, as before mentioned, are wont only to lick up fluid or soft bodies. Yet they only take the soft, more or less damp flour from germinating seeds, or from those the germination of which has been delayed, and which have again been dried, while they leave the hard dry flour from ordinary unsoftened grain. Moggridge substantiated this interesting fact by his attempts at artificial feeding. They only made an exception in favor of the fatty, oily remains of hemp seeds; these were gnawed on all sides, without being softened in water, while under ordinary circumstances the hard husks of the hemp seeds and of most other grains made this impossible. But if the husks burst owing to germination, the ants could then eat the softened and changed contents. In any case the peculiar parts of the ant's mouth are only fitted, as mentioned before, for taking liquid or soft substances; but they can very well scrape or scratch little bits of flour off seeds with their hard toothed mandibles.

Moggridge has also demonstrated that ants are subject to error and deception in spite of the instinct bestowed upon them by their creator, which was always to lead them aright, both in the choice of their nourishment and in so many matters of some of which mention has been made. It has already been said that they took china beads for grain. But less pardonable than this is it, that they took into their nests the little egg-like gall-apples of a small *Cynips* species (Gall wasp), which much resemble the seeds of the *Fumaria capreolata* (Fumitory) and put them in their granaries, clearly under the delusion that they were seeds. They are also apt to fall into error with respect to the weather, and it is quite untrue that they can foretell it, as Ebrard maintains. They turn back home and close the outlets when it rains, but that is all. Forel has often seen the *F. rufescens* and the *F. sanguinea* overtaken by heavy rain when out on a slave hunt, while he has observed others which were just setting out turn back if the sky was cloudy, and come out again if the sun won the day. They were not able to foresee that, in spite of the dark clouds, it would not rain! An army of Amazons was overtaken by a storm accompanied by sudden and heavy rain when thirty yards from their nest. They turned back without attaining their object, and arrived at their nest in great disorder when the chief mischief was done. On the other hand an expedition of the sanguine ants was, under similar circumstances, continued. Sudden storms surprise the ants, but those which approach slowly are discovered and avoided. The way in which an ant train returning from a plundering expedition, laden with stolen goods, may blunder as to its road back will be told later.

If the harvesting and storing of grain be the fashion among many species of ants in Southern Europe, this is naturally even more the case in hot or tropical lands. The activity of the Brazilian *Sa-uba* ants in this direction has already been spoken of. But in addition to these there are a good many other tropical and harvesting species, the number of which already discovered in the whole world Moggridge puts at nineteen. Dr. Delacour ("Rev. Zool.," May, 1848, 1849,) describes a giant ant of New Grenada, named *Arieros* by the natives, which emptied for him a whole sack of wheat during a single night. Lieutenant-Colonel Sykes speaks as follows of the Indian *Æcodoma*, or *Atta providens*, which does much harm in gardens and fields by its thefts of seeds ("Descr. of New Indian Ants" in "Trans. of the Ent. Soc.," 1836, p. 103): "In my morning walk I observed more than a score of little heaps of grass-seeds (*Panicum*) in several places on uncultivated land near the parade-ground; each heap contained about a handful. On examination I found they were raised by the above species of ant, hundreds of which were employed in bringing up the seeds to the surface from a store below; the grain had probably got wet at the setting-in of the monsoon, and the ants had taken advantage of the first sunny day to bring it up to dry. . . . Each ant was charged with a single seed; but as it was too weighty for many of them, and as the strongest had some difficulty in scaling the perpendicular sides of the cylindrical hole leading to the nest below, many were the falls of the weaker ants with their burdens from near the summit to the bottom. I observed they never relaxed their hold, and with a perseverance affording a useful lesson to humanity, steadily recommenced the ascent after each successive tumble, nor halted in their labor until they had crowned the summit, and lodged their burden on the common heap." Similar proceedings were often observed by Dr. Jerdon, ("Madras Journ. Lit. and Sc.," 1851, p. 46,) who had several packets of seeds he kept in his room emptied by these ants before he was aware of it. On the 7th November, 1866, Mr. Horne observed an Indian ant (*Pseudomyrma rufo-nigra*, Jerdon,) with the same habits near Mainpuri ("Hardwicke's Science Gossip," No. 89, p. 109). A long column of grain-carrying ants betrayed to him the way to a subterranean nest with five or six entrances, which he found in the middle of a floor of earth, about eighteen inches in diameter, firm, smooth, and kept perfectly clean. From this floor not less than thirteen roads led in all directions, which were perfectly smooth and clean, going over all the unevennesses of the ground, and these could be followed for from thirty to forty yards, until they were lost in the grass. Near the entrances were large rubbish heaps, placed carefully on one side, which consisted chiefly of the husks of the gathered corn.

In times of scarcity, observers say, that not only are the nests of these ants plundered, but the rubbish heaps are also seized, and are used as food in common with other grains. The time of year at which Horne made his observations was the beginning of the cold weather in November, *i.e.*, the beginning of the time of privation. Dr. Buchanan White sends from Capri ("Trans. of the Ent. Soc.," 1872, part 1,) under the 3rd of June, 1866, almost the same observation. "The perseverance with which a single ant would try and draw a pod four times his own length was very interesting; sometimes three or four ants would unite in carrying one burden. Near the formicarium was a great mass of *débris*, consisting of empty pods, twigs, emptied snail-shells, etc., cast out by the ants. The seeds appeared to be stored inside the nest, as in one that I opened the other day I found a large collection."

That harvesting-ants were to be found in Palestine is proved by the texts from old Jewish writings given in chap. v. Mr. F. Smith has lately confirmed the presence there of *Atta barbara*, and Mr. Moggridge has received accounts thence, which leave no doubt remaining, that the granaries of these ants in that country stretch over as wide an area as the largest of those observed by himself in Mentone.

But by far the most remarkable of these ants is found in Mexico. This is the *Myrmica* or *Atta malefaciens seu barbata*, or agricultural ant; it is a large brown ant which —incredible as it may sound—not only gathers corn, but also plants it and reaps it when ripe, so that practises a regular and complete system of agriculture, whereby, like a prudent husbandman, it makes suitable and appropriate arrangements for the different seasons of the year. Dr. Linecum and his daughters, without other help, watched this remarkable creature during ten years in the neighborhood of their home in Texas: and the famous Charles Darwin laid the observations made before the Linnæan Society in London. (See "Journ. of the Proceedings of the Linnæan Soc. of London," 1861, vol. vi., p. 29. Compare Buckley, "Proceedings of the Academy of Nat. Sc. of Philadelphia," 1860, p. 44): "The species which I have named 'Agricultural' is a large brownish ant. It lives in what may be termed paved cities, and like a thrifty, diligent, provident farmer, makes suitable and timely arrangements for the changing seasons. It is, in short, endowed with skill, ingenuity, and untiring patience, sufficient to enable it successfully to contend with the varying exigencies which it may have to encounter in the life-conflict. When it has selected a situation for its habitation, if on ordinary ground, it bores a hole, around which it raises the surface three and sometimes six inches, forming a low, circular mound, having a very gentle inclination from the centre to the outer border, which on an average is three or four feet from the entrance. But if the location is chosen on low, flat, wet land, liable to inundation, though the ground may be perfectly dry at the time the ant sets to work, it nevertheless elevates the mound, in the form of a pretty sharp cone, to the height of fifteen or twenty inches or more, and makes the entrance near the summit. Around the mound in either case the ant clears the ground of all obstructions, levels and smooths the surface to the distance of three or four feet from the gate of the city, giving the space the appearance of a handsome pavement, as it really is. Within this paved area not a blade of any green thing is allowed to grow, except a single species of grain-bearing grass. Having planted this crop in a circle around, and two or three feet from the centre of the mound, the insect tends and cultivates it with constant care, cutting away all other grasses and weeds that may spring up amongst it and all around outside of the farm-circle to the extent of one or two feet more. The cultivated grass grows luxuriantly, and produces a heavy crop of small, white, flinty seeds, which under the microscope very closely resemble ordinary rice. When ripe, it is carefully harvested, and carried by the workers, chaff and all, into the granary cells, where it is divested of the chaff, and packed away. The chaff is taken out and thrown beyond the limits of the paved area. During protracted wet weather, it sometimes happens that the provision stores become damp, and are liable to sprout and spoil. In this case, on the first fine day the ants bring out the damp and damaged grain, and expose it to the sun till it is dry, when they carry it back and pack away all the sound seeds, leaving those that had sprouted to waste.

" In a peach orchard not far from my house is a considerable elevation, on which is an extensive bed of rock. In the sand-beds overlying portions of this rock are fine castles of the Agricultural Ants, evidently very ancient. My observations on their manners and customs have been limited to the last twelve years, during which time the enclosure surrounding the orchard has prevented the approach of cattle to the ant-farms. The cities which are outside of the enclosure as well as those protected in it are, at the proper season, invariably planted with the ant-rice. The crop may accordingly always be seen springing up within the circle about the first of November every year. Of late years, however, since the number of farms and cattle has greatly increased, and the latter are eating off the grass much closer than formerly, thus preventing the ripening of the seeds, I notice that the 'Agricultural Ant' is placing its cities along the turn-rows in the fields, walks in gardens, inside about the gates, etc., where they can cultivate their farms without molestation from the cattle.

" There can be no doubt of the fact, that the particular species of grain-bearing grass mentioned above is intentionally planted. In farmer-like manner the ground upon which it stands is carfully divested of all other grasses and weeds during the time it is growing. When it is ripe the grain is taken care of, the dry stubble cut away and carried off, the paved area being left unencumbered until the ensuing autumn, when the same 'ant rice' reappears within the same circle, and receives the same agricultural attention as was bestowed upon the previous crop, and so on year after year, as I *know* to be the case, in all situations where the ants' settlements are protected from graminivorous animals."

Buckley also mentions that Dr. Linecum's daughter went daily into the garden to see the ants carry out their corn-stores, which often amounted to more than half a bushel.

CHAPTER IX.

CATTLE AND MILKING.

THUS has this little but wonderful creature, suiting itself to its life-conditions, reached a stage of culture to which man only attained after passing through two preceding long stages of hunting and shepherd life. But as though this were not enough, ants also exercise the usually accompanying duties of husbandry, care of cattle and milking, in a way that does as much credit to their taste as to their acuteness. They have as milch-cows—if not only, yet in preference to all other animals—the countless and easily obtained Aphides, which yield from their thick abdomens a sweet juice much liked by ants.* The ants are not alone in this kind of gluttony. Flies, wasps, bees, etc., are fond of this sweet liquid, and try to obtain it. Especially in autumn there is opportunity of seeing willow-trees quite covered with Aphides, accompanied by sucking ants and other insects. But none of these insects know how to manage the Aphides better than do the ants, which stroke their abdomens with their antennæ until they yield a drop of their sweet juice. This must be done in a soft and caressing manner peculiarly pleasant to the Aphides, for Darwin tried in vain to imitate the ants and to persuade them to yield their sweet excretion by stroking their abdomens with a fine hair. "I removed all the ants," says Darwin ("Origin of Species," ed. 1860, p. 210) "from a group of about a dozen Aphides on a dock plant, and prevented their attendance during several hours. After this interval, I felt sure that the Aphides would want to excrete. I watched them for some time through a lens, but not one excreted; I then tickled and stroked them with a hair in

---

* [The Aphides yield a liquid resembling honey both from the opening of the alimentary canal and from two setiform tubes. TR.]

the same manner, as well as I could, as the ants do with their antennæ; but not one excreted. Afterwards I allowed an ant to visit them, and it immediately seemed, by its eager way of running about, to be well aware what a rich flock it had discovered; it then began to play with its antennæ, on the abdomen first of one aphis and then of another; and each aphis, as soon as it felt the antennæ, immediately lifted up its abdomen and excreted a limpid drop of sweet juice, which was eagerly devoured by the ant." The behavior of ants towards Aphides has been known for a considerable time. Linnæus has called the Aphis the cow of the ants *(Aphis Formicarum vacca)* although he did not know that the ants took them within their dwellings, and there kept them as regular milch-cows. Huber alludes to this when he says: "An ant colony is the richer the more Aphides it keeps. They are its cattle, its cows, its goats. Who could have thought that the ants were a shepherd-nation!" The *Lasius brunneus*, or brown ant, which seldom leaves its nest, lives almost wholly, according to Forel, on the large bark lice, which it keeps and feeds in its rooms and passages generally hollowed out in the bark of trees. It manifests the greatest care for these animals, carries them away when the nest is uncovered, or, if they are too large to carry, leads them into the uninjured galleries. These lice have a very long proboscis, which they thrust deeply into the bark of the trees on the sap of which they live. They can only withdraw it with difficulty, and nothing is funnier to see than the indifference with which the ants compel the poor things to loose their hold by pulling and tearing at them, if the nest be opened or endangered. Although the drawing out of the proboscis can be managed if done slowly, yet there is danger of breaking it. *Lasius flavus*, the yellow ant, also lives entirely on the juice of the leaf, or rather root lice, which it keeps in the tree roots surrounding its nest. If its nest is uncovered, it carries away its beloved milch-cows with just the same care as it does its own larvæ —as Forel has often had an opportunity of observing. Many species, as already mentioned, build roofs and galleries of earth for them on plants and trees, so as to protect them from outside injury. Others collect the eggs of the Aphides in autumn, and thus bring them up and keep them in the interior of their own nests: but their love for their foster children does not prevent them from eating them up, skin and all, in times of need, when food is scarce. "They take as much care of these eggs," says Schmarda ("Intellectual Life of Animals," 1846), "as they do of their own." According to the same writer, they look so carefully after the safety of the Aphides that they make a kind of embankment of earth as a protection for the plants (especially spurge) on which they live. One day Huber found round the stem of a spurge a small dome or heap of earth, the smooth-walled interior of which gave shelter to a family of Aphides, which the ants would here milk undisturbed, protected from sun, rain, and strangers. The ants came and went to and fro from the neighboring ant-hills through a very small hole made in the lower part of the dome. Huber found a similar "stall" on a little poplar twig five feet from the ground. The leaf-scales of the common plantain, beneath which the Aphides living on this plant withdraw in August, when bloom and stalk are dried up, are walled up and covered in the ground with damp earth, so that plenty of room is left within which the milking can go on unhindered.

By their caresses and coaxings the ants are able to obtain from the Aphides a larger amount of their sweet juice than is otherwise habitually excreted. This is also the reason why trees and plants suffer which are much visited by ants. But the ants are not, as is generally thought, the direct, but the indirect, cause of the mischief, for first they largely increase the number of the Aphides by their care and tendance, and secondly, as the Aphides yield more to the ants, they must take more from the plant. If there be no ants present, they will of course do the same, but in smaller measure. They then throw their excrement out by a kind of jerk of the abdomen. But if ants are present they wait patiently till they come and relieve them of their burden. The drops are then seen to follow each other quickly, while otherwise the Aphides can remain quiet for a long time, excreting nothing. The ant takes as much juice as possible into the stomach, and can, as before mentioned, regurgitate the superfluity later for its comrades and larvæ.

The gall insects, parasitic on plants and trees, such as Kermes and Coccus, and especially the cochineal insect receive just the same attention from ants; and they, with the other plant lice, yield the ants, in our countries, the largest part of their food, although there is herein the greatest diversity between the different species, and although, as has been already said, the harvesting ants quite scorn the Aphides. Some species *(Leptothorax, Colobopsis)* feed directly on the juices of trees and plants; some are carnivorous, as *Pheidole, Tapinoma, Tetramorium,* and others, and prefer putrefying bodies, dead insects, and similar comestibles to sweet things. The gall insects are also kept by ants as milch cows within their nests, and McCook found Cocci as well as Aphides in the nests he examined; they were, however, kept in special, or separate rooms. The same observer saw the caterpillars of butterflies belonging to the genus *Lycæna* used as milch cows by a species of black ant (*F. subsericea*), since these caterpillars yield a kind of sweet liquid from minute glands in the abdomen.

If ants are seen running up and down the branches of a tree in great numbers, it is almost always only because Aphides are to be found on the tree. They visit fruit trees, for instance, only for the sake of the Aphides and gall-insects, and do not touch uninjured fruit. Many attempts have been made to destroy or hinder the clever little creatures in this proceeding, partly to test their intelligence, and partly to protect the trees from the injury inflicted on them. How difficult it is to do this will be seen immediately. They are indeed very easily frightened or startled away by anything strange and unaccustomed, but this only lasts until they have either recognised its harmlessness or learnt how to overcome the obstacle. For instance, if a ring of chalk be made round the trunk of a tree visited by ants, those which first come to it are startled and do not venture to cross the ring. But after a few of the braver have looked carefully into the matter, and have found that there is nothing to harm them, the rest follow them over the line. Professor Leuckart put a more difficult, and apparently an insuperable, obstacle in their way by spreading round the tree-trunk a broad band soaked in tobacco-water. When the ants coming from above arrived at the spot whereat this obstacle awaited them, they turned back and let themselves drop from the twigs of the tree and so arrived on the ground. But it was not to be so easily managed by the ants coming up from below. They convinced themselves that they could not cross the band without peril of life, and turned back. But Leuckart soon saw them coming back again, each carrying in its jaws a pellet of earth. This earth was placed on the tobacco-soaked barrier, and this was continued until a practicable road was made, on which the little creatures then ran backwards and forwards!

An exactly similar incident was observed by the famous Cardinal Fleury (1653-1743), a great admirer of ants, according to the information published by the distinguished French naturalist, Réaumur, in a fragment treating on ants, in his "Natural History of Insects" (1734-42). The Cardinal told Réaumur that he had smeared the trunk of a tree with a ring of birdlime to keep back the ants, and that the ants had surmounted the difficulty by bringing up little pellets of earth, small stones, etc. The same observer saw a number of ants bring little pieces of wood and make therewith a bridge, in order to cross an obstacle put in their way, a vessel of water surrounding the bottom of an orange tub.

The ants behaved in yet more ingenious fashion under the following very similar circumstances; Herr G. Theuerkauf, the painter (Wasserthorstr. 49, Berlin), writes to the author, November 18, 1875: "A maple tree standing on the ground of the manufacturer, Vollbaum, of Elbing (now of Dantzic) swarmed with Aphides and ants. In order to check the mischief, the proprietor smeared about a foot-width of the ground round the tree with tar. The first ants who wanted to cross naturally stuck fast. But what did the next? They turned back to the tree and carried down Aphides which they stuck down on the tar one after another until they had made a bridge over which they could cross the tar-ring without danger. The above named merchant, Vollbaum, is the guarantor of this story, which I received from his own mouth on the very spot whereat it occurred."

Apart from the very ingenious surmounting of the obstacle, what becomes of that innate affection and love for Aphides, which the instinct-philosophers are obliged to ascribe to ants, since, without thought or sympathy for their beloved fosterchildren, they sacrificed them to a dreadful death, when a higher object seemed at stake!

For a scarcely less interesting observation than the

above on the relations between ants and Aphides, and the thereby developed intelligence of the former, the author is debtor to Herr Nottebohm, inspector of buildings, at Karlsruhe, who related the following on May 24th, 1876, under the title, " Ants as founders of Aphides' colonies : " " Of two equally strong young weeping-ashes, which I planted in my garden at Kattowitz, in Upper Silesia, one succeeded well, and in about five or six years showed full foliage, while the other regularly every year was covered, when it began to bud, with millions of Aphides, which destroyed the young leaves and sprouts, and thus completely delayed the development of the tree. As I perceived that the only reason for this was the action of the Aphides, I determined to destroy them utterly. So in the March of the following year I took the trouble to clean and wash every bough, sprig, and bud before the bursting of the latter, with the greatest care by means of a syringe. The result was that the tree developed perfectly healthy and vigorous leaves and young shoots, and remained quite free from the Aphides until the end of May or the beginning of June. My joy was of short duration. One fine sunny morning I saw a surprising number of ants running quickly up and down the trunk of the tree ; this aroused my attention, and led me to look more closely. To my great astonishment I then saw that many troops of ants were busied in carrying single Aphides up the stem to the top, and that in this way many of the lower leaves had been planted with colonies of Aphides. After some weeks the evil was as great as ever. The tree stood alone on the grass plot, and offered the only situation for an Aphides' colony for the countless ants there present. I had destroyed this colony; but the ants replanted it, by bringing new colonists from distant branches and setting them on the young leaves."

Where nature, therefore, does not voluntarily provide for the coveted presence of their beloved milch-cows, the ants know how to take the trouble on their own shoulders in a suitable place. Sometimes they attain their object in a shorter but more perilous way, by simply fighting with their relations or rivals for possession of the Aphides, just as we have seen similar conduct among harvesting ants, and shall see yet more among other genera. Forel saw a colony of *Formica exsecta*, which he brought with him from Mont

Tendre and established on the edge of a little wood near Vaux, at once attack without fear two nests of *Lasius niger* and *Lasius flavus*. After they had slain many of their enemies, they rushed upon the shrubs growing around, and chased off the ants which were on them in order to seize their Aphides. They tried to do the same with an oak occupied by *Camponotus ligniperdus*, whereon these large, strong, and warlike ants kept their Aphides. They made unheard-of efforts to obtain their object, but failed. They were flung back by their terrible foe and killed by hundreds, and at last gave up the enterprise. They indemnified themselves therefor by seizing a number of grasshoppers' holes, and chasing away the inhabitants. As Forel remarks, this is a habitual method of obtaining for themselves a temporary dwelling adopted by almost all kinds of ants. Three or four of them push into a grasshopper's hole; it comes out, and tries to drive away its foes by biting and seizing them. But the ants fling themselves upon it, hold down its legs and sprinkle it with their poison. The grasshopper yields, leaves its nest, and is fortunate if it escapes with life. Its dwelling is then taken possession of by the ants. Ants fight over sugar thrown in their way, just as they do over Aphides. Forel one day threw a piece of sugar between two columns of the *Lasius emarginatus* and the *Tetramorium cæspitum* (turf-ant) which had been fighting, but were then retreating on account of the sun. The battle broke out anew, and that most vigorously. The *emarginati* were beaten and pursued to their nest, while the victors seized the sugar. In South America there is an ant, *Myrmica* or *Atta saccharum*, which lives only on sugar, or rather on the juice of the sugar-cane, and commits terrible depredations in sugar plantations by burrowing under the roots. It is thence called the sugar-ant.

McCook (*loc. cit.*, p. 275) gives a very remarkable observation as to the milking of Aphides—made on the American mound-building ant (*F. exsectoides*), whose large buildings have already been described—after he had very accurately described the process of milking. He noticed that of the workers returning to the nest from the tree on which the milking was going on, a far smaller number had distended abdomens than among those descending the tree itself. A closer investigation showed that at the roots of the trees, at the outlets of the subterranean galleries, a number of ants were assembled, which were fed by the returning ants after the fashion already described in feeding the larvæ, and which were distinguished by the observer as "pensioners." McCook often observed the same fact later, among, with others, the already described Pennsylvanian wood-ant. Distinguished individuals in the body-guard of the queen were fed in like fashion. McCook is inclined to think that the reason of this proceeding is to be found in the "division of labor" so general in the Ant Republic, and that the members of the community which are employed in building and working within the nest, leave to the others the care of providing food for themselves as well as for the younger and helpless members; they thus have a claim to receive from time to time a reciprocal toll of gratitude, and take it, as is shown very clearly, in a way demanded by the welfare of the community.

The plant-lice and gall-insects are not, as said above, the only milch-cows of the ants. There is a large number found in the nests with the other frequently seen insects, which are all comprised under the name of *Myrmecophila*, or ant-friends, and which differ very much according to varying circumstances, in different species and in different lands. All myrmecophilous insects are alike in this, that they secrete a sweet juice which the ants can feed upon. In order to obtain this object the ants spare no time and no trouble, and they treat their sweet friends with a love, a thoughtfulness and a care which would be worthy surprise as an image of friendship, were it not for the strong egoistical interest mixed therewith. Most of the Myrmecophila are blind, because living in the perpetual darkness of the nest they have no need of eyes, or rather because their visual organs have gradually become rudimentary by continued disuse. They are therefore, being unable to seek food for themselves, entirely dependent on their lords and protectors the ants, just in the same way as, and still more than, domestic animals, such as the dog, are dependent upon man. Unfortunately most of the Myrmecophila—called also ant parasites, and reckoned by Lespès as containing three hundred different species—are still very little known. Especially remarkable, according to Lespès, is the behavior of ants towards a kind of blind beetle, the Claviger beetle, which has large antennæ in lieu of eyes, and only rudimentary wings. Its movements are very slow, and the parts of its mouth are only suitable for fluid food. It cannot feed itself, but is fed and nourished by the ants just in the same way as they feed each other, from mouth to mouth. In return the beetle yields each time a very pleasant-flavored juice: the ants lick this up, pressing the juice-yielding part with their jaws in every possible, but non-injurious fashion. As soon as an ant has given a Claviger anything to eat, or rather to drink, after a previous communication with their feelers, it indemnifies itself by sucking at its body. This whole proceeding is clearly a sort of refined gluttony, for it would have no sense merely as a matter of nourishment, as the ant must give as much, or more, food material to the beetle than the latter is able to return. And this is the more likely as the ants are always regarded as gluttons. If, for instance, honey is given to them, of which they are very fond, they will leave everything in the lurch, even their larvæ, in order to swallow as much as possible, while they will not do this with other things they like less, as the juices of insects.

The Claviger species always stay in the nest. A seeing Staphylinus, a large species of beetle *(Lomechusa; Atemeles)* which has been observed by Lespès, leads a very changeful life. It has fully developed wings and flies about outside the nest during the greater part of the day. But it is unable to procure its own food, and when pressed by hunger returns to the nest to be fed by the ants. Lespès has seen one of these approach an ant, and communicate its want by touching it with its feelers. When the meal was over, the well-bred Staphylinus offered its abdomen to its feeder, so as to discharge its debt of obligation. These beetles behave just like naughty boys, who are always playing about outside, and only come home when meal-time is near. Their outer appearance in no fashion betrays that they are so clumsy and helpless as not even to be able to find their own food.

E. Schröder ("Zoolog. Gardens," 1867, p. 227) saw a specimen of the *Lomechusa strumosa* fed and milked by red-brown ants, the body of the beetle manifesting a quivering movement of delight so long as the stroking lasted. Another species of Staphylinus (*Myrmedonia*) is an enemy of the ants, but only ventures into their nests during the winter, when they are half frozen. In summer it would soon be torn to pieces. As a rule, it lies in wait near the roads frequented by ants and seizes solitary passers-by, tearing open and devouring the abdomen filled with sweet juice. Forel regards all Myrmecophila—of which Taschenberg reckons three hundred species in Germany alone—as direct or indirect parasites, and as only accidental parts of the economy of the ant polity. He has found a large number of them, especially of species of beetles, in the ants' nests, but could not establish their exact relations with the ants. That these beetles are not always friends of the ants is proved by the following. Forel saw a tolerably large beetle (*Hister quadrimaculatus* L.) appear in the midst of a number of *F. pratensis*, returning from a battle with the red ants, and plunge his head into a pupa. The ants rushed furiously upon him, covered him with bites and poison, and tried to tear away his prey. But the hard outer case of the beetle made all their efforts useless, and the beetle appeared to be perfectly conscious of this and to be sure of his work, so did not let himself be terrified. He kept his head and forefeet firmly fixed to the pupa, and used his remaining four legs for retreat, and so came off uninjured with his prey. Another species of *Hister*, which tried a similar attack on *F. cæspitium* (turf ants), was seen by Forel to expire under their stings.

In addition to the beetles, Forel found in considerable numbers in the ants' nests a large, long, white, ringed larva, which was fed and tended by the ants like their own larvæ, and which he regards as the larva of an unknown beetle, which becomes a pupa and emerges elsewhere. Perhaps the ants confound this larva with their own, but some deeper fact may lie in it.

Moggridge very often found in the nests of the harvesting ants—together with little white spring tails, or *Poduri*, the " silver fish " (*Lepisma*), and a small species of wood louse (*Coluocera attæ*)—the larvæ of an elater-beetle, of which the ants appear to take great care. Moggridge thinks that this care is quite selfish, and that it is shown in order to make use of the tunnels of the larvæ. A little cricket (*Grillus myrmecophilous*) is also found in some ants' nests in Italy and France.

Still less is known of the Myrmecophila of non-European

lands than of European, although they are not less numerous. Bates found a peculiar species of snake (*Amphisbœna*) in the nests of the *Sa-uba* ants. Julius Fröbel saw in Mexico an ant-colony changing its dwelling. In the procession marched some little beetles, resembling our *Coccionella semipunctata*. If one of these tries to get out of the line of march it is quickly brought back by the ants at its side ("From America," Leipzig, 1875, I. 275). "In Brazil the place of the Aphides is taken by the larvæ and nymphæ of certain crickets, namely, of *Cercopis* and *Membracis*, which sit on the plant-stalks sucking at the sap, and from time to time excrete from the abdomen a drop of sweet juice which is eagerly licked up by an ant (*F. attelaboides*), which caresses the cricket and helps it in changing its skin, just as our ants do with the Aphides. When Aphides, which had not previously been there, were introduced into the gardens of Rio Janeiro, the ants soon recognised their useful properties" (Perty, on "The Intellectual Life of Animals," 2nd ed., p. 315). "According to Audubon certain leaf bugs are used as slaves by the ants in the Brazilian forests. When these ants want to bring home the leaves which they have bitten off the trees, they do it by means of a column of these bugs, which go in pairs, kept in order on either side by accompanying ants. They compel stragglers to re-enter the ranks, and laggards to keep up by biting them. After the work is done the bugs are shut up within the colony and scantily fed" (Perty, pp. 329 and 330).

A species of ant which lives in the same country as the remarkable agricultural species, has carried farthest the art of cattle-keeping and milking. It is the *Myrmecocystus mexicanus*, discovered thirty years ago by a Belgian naturalist, Wesmaël. Among these certain neuter ants, which must belong to a special caste, take the place of the Aphides and Myrmecophila, and so fill their very dilatable abdomens with honey that they look like little round bottles, and can be brought to market as an article of commerce.

"These ants," says Blanchard, "which are very numerous round the town of Dolores and are known in the country under the name of *Basileras*, live in subterranean dwellings, which do not betray their presence outside. In the earlier part of their life they have an abdomen of the usual size. But it gradually enlarges enormously, owing to the collection of a syruplike liquid therein, until it resembles a transparent flask. In this condition these so-called honey-ants are unable to move, and hang motionless from the roofs of their dwelling. The women and children of the neighborhood dig up the nests and suck the *Basileras*. If they are brought to table, the head and thorax are pulled off, and the little honey-filled bladder laid on a plate."

These honey-ants are fed in a special way by their neuter sisters, and are then milked. They never leave the nest, and are therefore in the full sense of the word, like the Clavigers, "stall-cows." Here also the gluttony of the ants and their excessive fondness for honeylike sweets apparently play the chief part.

According to Dr. C. Crüger ("Journal of the Union of Naturalists," Hamburg, II. vol., 1876) the whole community appears to be made up of three kinds of animals, perhaps of different species, one consisting of the feeders and guardians of the honey-preparing caste which never leave the nest, and which bring them pollen and petals with nectaries, while a third large and strong kind with very powerful jaws act as soldiers, and have the duty of guarding the home. They place themselves in a double row in front of the nest, patrol in all directions, and only step out of their ranks to kill an approaching enemy or stranger. There are thus carters, merchants, and soldiers.

CHAPTER X.

INTELLIGENCE AND LANGUAGE.

THEIR love for honey makes ants very dangerous enemies of beehives, into which they often win their way in the most cunning and subtle fashion. Karl Vogt relates in his "Animal Societies" a story, since become very well-known, of the apiary of a friend which was invaded by ants. To make this impossible for the future, the four legs of the beehive-stand were put into small, shallow bowls, filled with water, as is often done with food in ant-infested places. The ants soon found a way out of this, or rather a way into their beloved honey, and that over an iron staple with which the stand was attached to a neighboring wall. The staple was removed, but the ants did not allow themselves to be defeated. They climbed into some linden trees standing near, the branches of which hung over the stand, and then dropped upon it from the branches, doing just the same as their comrades do with respect to food surrounded by water, when they drop upon it from the ceiling of the room. In order to make this impossible, the boughs were cut away. But once more the ants were found in the stand, and closer investigation showed that one of the bowls was dried up, and that a crowd of ants had gathered in it. But they found themselves puzzled how to go on with their robbery, for the leg did not, by chance, rest on the bottom of the bowl, but was about half an inch from it. The ants were seen rapidly touching each other with their antennæ, or carrying on a consultation, until at last a rather larger ant came forward, and put an end to the difficulty. It rose to its full height on its hind legs, and struggled until at last it seized a rather projecting splinter of the wooden leg, and managed to take hold of it. As soon as this was done other ants ran on to it, strengthened the hold by clinging, and so made a little living bridge, over which the others could easily pass. The great black ants of the East Indies were seen by Sykes to act in a yet more ingenious way. "In Sykes' house the dessert was placed on a table in a locked verandah, and covered with a cloth, the legs being placed in vessels of water. But the ants waded through, or, if the water were too deep, clung to each other with their strong legs, and so reached the feet of the table and the Chinese sweetmeats, and although hundreds were killed every day, the next day saw new crowds. Sykes then surrounded the legs of the table with a ring of turpentine, but after a few days they were again among the sweet fruits. The edge of the table stood about an inch from the wall. The large ants clung to the wall with their hind legs and stretched the fore legs across to the edge of the table, and so many managed to cross. Sykes pulled the table further away from the wall, but they then climbed up the wall to about a foot above the table, gave a spring and fell on the fruits" (Perty, p. 341).

The ants are as greedy about syrup and syrup-like fluids as they are about sugar, honey, and sweet fruits. Their cleverness in discovering these is so great, that as the instinct-mongers say, their instinct "borders on human reason;" in reality it is reason, and often surpasses the human acuteness which it is thus vainly sought to defend. When an ant has found such a treasure, it first obeys the inflexible law of egoism, and fills its own stomach as much as possible, until it is quite swollen. It then remembers its duties to its fellowmen, or rather fellow-ants, and after it has left the place, returns in a short time with a number of its comrades, which now do the same.

Dr. Franklin (cited in "Bingley," iv., p. 176,) tells how, in order to test the intelligence of the ants, he put a little earthen pot filled with treacle into an out-of-the-way closet. The ants soon appeared in crowds, and devoured the treacle. He then drove them away, and hung the pot from the ceiling by a string, so that, as he thought, no ant could reach it. A single ant was accidentally left behind in the pot. It eat as much treacle as it could, and then wanted to get away. After long searching it found the string, and made its way back along it. By way of the ceiling and down the wall it again reached the ground. But it had hardly been away an hour when a large swarm of ants arrived, climbed up the wall, ran along the ceiling to the string, which led them to the treacle-pot. They repeated this manœuvre, relieving each other in troops, until all the treacle had disappeared.

Such a proceeding, one among hundreds of similar observations, necessarily suggests two questions, the investigation of which cannot be avoided in a psychological view of animals:

First; How do the ants find their way to a place very distant from their nest at which a source of nourishment has been discovered, and which they are unable to see?

Second; By what means do they communicate to each other the discovery of such a treasure and induce their comrades to follow them to such a place?

As to the first question, there can be no doubt that its answer lies in the excessively fine scent of the ants, while their sight appears to be rather weak. The first is so sensitive, that the approach of a human hand to an ant road is enough to startle the prudent little creatures. Lespès says that if a hand is laid for a moment on an ant path while it is clear and then taken away again, the first ant that comes along starts back terrified as soon as it comes to the spot touched, and generally runs away as fast as possible. A second comes, and a third, and all act in the same way. At last one appears which either has a less keen scent or is less timid, and which passes the barrier. As soon as this has been done without danger, all the rest follow. Forel made just the same observation with *Lasius emarginatus*. If a finger is put for a moment on its road when none is there, the first ant that comes up stops suddenly, stretches its feelers in the air and turns back. Soon others come up in a clearly uneasy way, run up, search over the place, and do not pass by the spot until they are convinced that there is no danger. Forel says further that nothing is more necessary in scientific ant observations than to keep the mouth and nose covered with the hand, for that the lightest human breath is sufficient to frighten the ants and to cause the failure of the experiment. He kept an ant-colony (*Strongylognathus testaceus*) with a jar of honey in a cage, or vivarium, surrounded by a high and thick plaster wall. He then ceased feeding another ant-colony (*Camponotus herculaneus*) in the same room, so as to induce them to go out. The *Camponotus* ran about everywhere, and by the help of their scent soon discovered the honey, although they could not see it, nor climb over the wall. They tunnelled through the gypsum, which Forel vainly repaired, and stole the honey until the wall was made so strong that they could not get through it. Only a single ant got through, in what way is unknown, and so crammed itself with honey that it could not return. Forel found it the next morning leaning against the inner side of the wall, and unable either to make a tunnel through or to climb over. Thus we have an ant so lacking in the " instinct " of feeding itself that it was unable to conquer its greediness! Jars containing jams or sweetmeats, placed in water, are regularly blocked up with swarms of ants, which have scented the sweetness and in vain seek ways and means of penetrating to it.

As to the searching for or finding again a road once passed over by them, this is all the easier to them, owing to the fairly strong scent which they are known to leave behind them.

In that which concerns the second part, the power of communication, or language, the very sensitive antennæ play the chief part, being well supplied with nerves. Two ants, which are communicating with, or speaking to each other, stand face to face, and mutually tap each other in the most rapid manner with their peculiarly motile feelers, striking each other's heads, and so on. That they are thus able to impart quite detailed information, and as to distinct things, is proved by countless examples, some of which have already been mentioned. Jesse writes : " I have often put a small green caterpillar near an ant's nest; you may see it immediately seized by one of the ants, who after several ineffectual efforts to drag it to its nest, will quit it, go up to another ant, and they will appear to hold a conversation together by means of their antennæ, after which they will return together to the caterpillar, and, by their united efforts, drag it where they wish to deposit it. I have also frequently observed two ants meeting on their path across a gravel-walk, one going from and the other returning to the nest. They will stop, touch each other's antennæ, and appear to hold a conversation ; and I could almost fancy that one was communicating to the other the best place for foraging " (" Gleanings," vol. 1. pp. 18, 19, ed. 1838).

In a letter to Charles Darwin, Hague (quoted by Landois,

"Animal Language," 1874, p. 129) relates that he one day killed with his finger a number of ants which daily came out of a hole to some flowers standing on a chimney-piece and did not allow themselves to be disturbed by brushing their path. The result of this was that all new comers at once turned back again and tried to persuade those of their companions not yet conscious of the danger to do the same. Those meeting each other held a short conversation, the result of which was not an immediate return, the arriving ants seeking a proof for themselves. When the war-making ants are going to take the field, they hold council in the same way, as will later be more fully explained, and talk over the resolution arrived at. When a hungry ant wants food, it communicates with its comrades by movements of its feelers. The helpless larvæ are in the same manner bidden to open their mouths for food. Mutual likes and dislikes are also signified by means of this gesture-language. Landois is further of opinion that, judging by his own observations, ants must have not only a gesture-language but also a spoken or voice language, even though it be not audible to human ears, while Sir John Lubbock (Journ. Linn. Soc. Zool. xiv., p. 265 *seq.*) thinks himself obliged to deny this, following his own investigations. For instance Landois threw a large live cross-spider on a very populous ant-heap. In a trice the whole colony was alarmed, and that with a celerity which only seemed explicable to him by some kind of *acoustic* communication. A large number of ants flung themselves on the spider, and a tremendous struggle ensued which ended in the complete defeat of the intruder. Landois also succeeded in proving the existence of a sound apparatus, or strident (rasping) organ in the abdomen of ants. According to Ponera the strident noise can be heard by the human ear, but not by the ants themselves.

Further, the language—or the power of communication—must vary in richness and completeness in different species. For it has already been mentioned, for example, that on the occasion of a projected change of dwelling one ant will take another between its jaws and carry it to the place selected for the new abode, while others do not need so practical a method of communication, but understand each other by signs and gestures. The most frequent, or one of the most frequent occasions for mutual communication and gathering together may well be in the case of the discovery of a food supply as in the incidents mentioned above. Lubbock (*loc. cit.*) noticed how an ant, which had found such a treasure, brought some of its friends, and how these informed others, and so on; but all kinds did not behave in this fashion, and some (such as specimens of *F. fusca*) only sought the food for their own consumption all day long. The latter the more resemble us men, who generally keep as secret as possible discoveries of places of food supply! The same observer was able to place it beyond doubt, by many ingenious experiments, that, as maintained above, their communications are not only of a general kind, but can deal with quite concrete things and directions; for instance, that a place with many larvæ will, in pursuance of the invitation of the ant-guide, be sought by a larger number of ants than will one with few. Thus planned deceptions have been shattered by the wisdom of these little creatures. The author is indebted to the already mentioned and excellent letters of Dr. Ellendorf, of Wiedenbrück, for many interesting observations on the power of communication among ants. Dr. Ellendorf writes :—

" I lived for a long time on the island of Omotepe in the lake of Nicaragua and had there an opportunity of watching these little creatures daily and hourly under compulsion, for they were inmates of my house. I had scarcely built my rancho when they were to be found therein, at first singly, then more and more, busily hunting over every corner. It was soon plain to me how fully they possessed the power of communicating with each other. I had hung the skin of a bird out to dry, and the next morning I only found a little heap of feathers on the ground. About 180 yards off I found the nest of the ants which had destroyed the skin, and which had clearly been summoned by a comrade which had discovered the skin and the way to it. In order to watch this proceeding more closely I fastened near my table a stick with a cross piece, and hung a dead bird from it by a string. Before long I saw an ant run down the string to the corpse. It walked deliberately over it, stood still here and there, and carefully tried it all round with its feelers. After thus busying itself for perhaps a minute, it ran up the string again and down the stick to the floor. As soon as it reached the ground it ran quickly about as though seeking someone. Suddenly it met a sister, and both stood still. It touched the latter with its antennæ, and then ran in the same way to a second and third and slipped out through the bamboos, so that I could no longer follow it. It was not long before an ant came down the string, then a second, a third, and so on. In an hour and a-half the body was thoroughly perforated by them, and in twenty-four hours all the flesh had disappeared off the bones. The next day I repeated the same manœuvre, but laid a piece of sugar on the ground near the stick, so as to see if the first who might discover the dead body would tell his comrades which were employed with the sugar. An ant soon ran down the string to the bird. It tried it, and then ran back to the ground. After running about for a moment, it mingled with the rogues busy with the sugar. Here I of course lost sight of it, but I noticed a great commotion in the crowd. About half left the sugar, and in a short time the dead body was again covered with ants. It was clear that the finder had informed the others!

"The next day I hung a dead body by a string tó a crossbeam of the rancho, so that it dangled freely in the air. Until the evening not an ant was to be seen, but the body was half-devoured by the next morning. So during the night an ant out hunting had discovered the body and had communicated the fact to the others.

"It is a hard matter to protect any eatables from these creatures, let the custody be ever so close. The legs of cupboards and tables in or on which eatables are kept are placed in vessels of water. I myself did this, but I none the less found thousands of ants in the cupboard next morning. It was a puzzle to me how they crossed the water, but the puzzle was soon solved. For I found a straw in one of the saucers, which lay obliquely across the edge of the pan and touched the leg of the press: this they had used for a bridge. Hundreds were drowned in the water, apparently because disorder had reigned at first, those coming down with booty meeting those going up. But now there was perfect order; the descending stream used one side of the straw, the ascending the other. I now pushed the straw about an inch away from the cupboard leg; a terrible confusion arose. In a moment the leg immediately over the water was covered with hundreds of ants, feeling for the bridge in every direction with their antennæ, running back again and coming in ever larger swarms, as though they had communicated to their comrades within the cupboard the fearful misfortune that had taken place. Meanwhile the new comers continued to run along the straw, and not finding the leg of the cupboard the greatest perplexity arose. They hurried round the edge of the pan, and soon found out where the fault lay. With united forces they quickly pulled and pushed at the straw, until it again came into contact with the wood, and the communication was again restored."

To this we may add a very interesting observation of the Franciscan father, Vincent Gredler, of Botzen, recorded as follows in " Zool. Gardens," xv. p. 434. "In Herr Gredler's monastery one of the monks had been accustomed for some months to put food regularly on his window-sill for ants coming up from the garden. In consequence of Herr Gredler's communications he took it into his head to put the bait for the ants, pounded sugar, into an old inkstand, and hung this up by a string to the cross piece of his window and left it hanging freely. A few ants were in with the bait. These soon found their road out over the string with their grains of sugar, and so their way back to their friends. Before long a procession was arranged on the new road from the window-sill along the string to the spot where the sugar was, and so things went on for two days, nothing fresh occurring. But one day the procession stopped at the old feeding-place on the window-sill, and took the food thence, without going up to the pendant sugar-jar. Closer observation revealed that about a dozen of the rogues were in the jar above, and were busily and unwearyingly carrying the grains of sugar to the edge of the pot, and throwing them over to their comrades down below!"

We here see an action exactly similar to that we found among harvesting ants, some of which loosened seeds from the stalks of plants and shook them down, while others below gathered them up.

CHAPTER XI.

SLAVERY.

EVERYTHING which has hitherto been related about ants, their behavior and their characteristic qualities, their polity, their buildings and road-making, their storing-up of provisions, their agricultural, malting, cattle-tending, and milking proceedings, their acuteness in foraging for food, their division of labor, their power of inter-communication, etc., is indeed most noteworthy, and is fitted to awaken in us respect and admiration towards the little creatures. But all that has been told retires into the background as to psychological significance in the intellectual life of these animals, when we find or remember that the ants, as already said, for an unknown length of time have had a politico-social institution which had played and still plays a great part in the history of human nations and civilization. This institution, indeed, seems at first sight to harmonise badly with the otherwise social-democratic tendency and arrangements of the Ant Republic. But when we remember that slavery existed in the republics of antiquity and not only well agreed with the rest of the polity, but was even an essential support of the same, we can scarcely deprive the Ant Republic of its democratic character on account of slavery. And this the rather since slavery among ants is as mild, if not milder, than it was in Greece and Rome, where freed slaves were often known to rise to the highest offices and dignities of the state, where as in Rome, Greek slaves were the tutors and trainers of the young, and where slavery, odious as it may be in and for itself, none the less contributed to the general advance of civilization. Besides slavery among ants is on a very important point far superior to that of man, and it may be said without question that in this respect ants think and act more humanely than men themselves! For instance, they never allow grown-up members of their race, come to their full antly consciousness, to be enslaved, whereas human slave-makers are known never to have the smallest scruple on this head. For the ant-kidnappers only steal larvæ and pupæ, which they bring up as regular slaves within their dwellings, so that these last have never tasted the condition and the sweetness of freedom. Only quite young ants, one or two days old, recognizable by their clear color, which are not yet out of long-clothes and do not yet know what is " manly (or womanly) pride before the throne of a king," are tolerably often made into slaves, and accustom themselves quickly and easily to their new position. The slaves of the ants, however, do not seem to be conscious of the loss, or rather of the absence of freedom, and as a rule work willingly and uncompelled in common with their masters at all the tasks necessary for the maintenance of the colony, such as building the dwelling, searching for plant-lice, tendance and feeding of larvæ and pupæ, and so on, and even fight against members of their own species in company with their robber-lords. They are regarded more as friends, brothers, or helpers than as real slaves. They never think of escaping from slavery by flight, although Forel, as will be told afterwards, once observed a revolt among them. This rule applies at least to the Swiss species observed by Huber and A., while in the south of England colonies have been seen in which the slaves never leave, or venture to leave the nest, and are thus, in the true sense of the word, domestic slaves.

As regards the slave holders themselves, we are acquainted with three kinds in Europe *(F. rufescens, F. sanguinea,* and *Strongylognathus)*, of which only the two first are accurately known. The most interesting among them is the often mentioned and famous Amazon *(Formica* or *Polyergus rufescens)* [the rufescent ant of Kirby, TR.], whose noteworthy doings and habits were first exactly observed and described by Huber. It is a large, strong, very active, shining-red ant, which behaves just as do human lords as a rule ; namely, it does not work, but leaves everything to be looked after by its servants, slaves, and laborers. It does not even eat by itself, but lets itself be fed by its slaves, and behaves just like the Llama of Thibet, who also has food put into his mouth by his slaves, because it is thought beneath the dignity of so great a lord to serve himself. Do the Amazons think the same? or do they follow the maxim held by many men that work is shameful? No—the ant has a far better ground for its conduct, or a far better and more cogent excuse than its human examples or followers. It *cannot* feed itself, and cannot discharge the regular duties of an ant, owing to its long, narrow, and strong mandibles, which do not form a toothed edge, as in other species, but run out into a sharp and strong point, so that they must be regarded as regular jaws. These jaws are indeed admirable for use as weapons, and are specially adapted to pierce the head and brain of an enemy, but make self-feeding and work impossible. The Amazon is therefore entirely dependent on the goodwill of its slaves. Without their help it would starve to death, and the whole colony would perish from want of care and nourishment.

Huber placed a number of ants (about 30) with their larvæ and pupæ and some earth in a box, and supplied them with plenty of food. After the lapse of only two days some of the ants died of hunger, or rather of thirst, although, according to Forel's experience, ants can live for four weeks without food, if the air or the earth is sufficiently damp. The Amazons were neither able to feed themselves, nor tend their young, nor work in the earth. Huber then put in a single ant of the slave species, and it quickly put everything to rights. It fed young and old with the honey lying there, began to build cells for the pupæ and larvæ, cleaned them and so on. In order to substantiate this observation Lespès one day put a piece of moistened sugar in front of a nest of Amazons. It was soon discovered by one of the ants of the slave-species (*F. fusca*, or small black ant [negro, Tr.]). It eat as much as it could and then went back into the nest. Other gluttons soon appeared, and the delicious meal was busily devoured. At last Lespès noticed some Amazons come out. They first ran about in a puzzled way, without touching the sugar, until at last they began to pull at the legs of their duty-forgetting slaves, thereby reminding them that they should be served. This was done and all parties seemed content.

Forel also never saw an Amazon eat alone. If one was hungry, it tapped a slave on the head with its feelers, until the latter brought up or regurgitated a drop of nourishment from its proventriculus and gave it to its master, mouth to mouth. All the other relations of Huber were fully confirmed by Forel. He put twelve Amazons with pupæ, larvæ, and abundant food (dead insects, larvæ, meat, honey, sugar) in a glass jar filled with damp earth. The Amazons did not move about but sat together in a corner. When Forel compelled one to go to the honey it behaved very clumsily, messed its forefeet and its feelers in it, and turned back again to its corner. Its whole conduct was exactly the opposite of what ants are accustomed to do under similar circumstances. The other Amazons also all avoided the honey, instead of touching it. On the other hand they several times vainly begged to each other for food. This lasted for several days, and all the food remained untouched. Two of the Amazons died, the others remained in good health, thanks to the damp earth. The larvæ visibly fell away. As soon as an Amazon came near them they turned to it to ask for food, but it contented itself with gently touching them with its antennæ, apparently in order to explain that it was unable to do as they wished. After the lapse of seven days all remained in the same state ; Forel then took away the meat and the dead insects, which had putrefied, and this time brought a slave belonging to the *F. rufibarbis* or *cunicularia*, a sub-species of *F. fusca*. The new comer was immediately surrounded by the Amazons, begging for food. At first it pushed back the crowd. But when it discovered the honey it filled itself in less than ten minutes, and then began to feed the Amazons, one after another. Without moving, it let a clear drop, as large as the head of an Amazon, run out of its mouth, and this was at once swallowed by the Amazon, which caressed its friend with its antennæ and fore-feet. The slave eat up the store of honey in this way, and divided it among its masters. It then began to look to the pupæ, all the larvæ having perished for want of nourishment. Another day Forel gave it a companion, and the two together now built rooms for the pupæ and for their masters. W. M. R. ("Nature," 1879, No. 1) made an exactly similar observation. He put thirty Amazons with four larvæ and some honey into a box, and after two days found five of them dead. He then put in three slaves, which at once set to work, fed the Amazons, acted as nurses to deliver two larvæ which were just emerging, and had brought everything to order in a few hours. The observer never saw the lazy little things feed themselves. Lubbock also could only keep alive a captive *Polyergus* by placing with it daily, during an hour, a slave which cared for and fed it.

All this sufficiently shows how thoroughly dependent the Amazons are upon their slaves. They also sometimes allow themselves to be carried by their slaves, although the latter are smaller and weaker. But this does not prevent the Amazons from picking up the slaves and carrying them off when changing their nest, or when otherwise necessary, for they very well know that they cannot live without them. Huber watched such an occurrence, when an army of Amazons found a deserted nest of *F. fusca*, and emigrated to it, each Amazon seizing a slave and carrying it off. All domestic duties are completely handed over to the slaves, while the masters give themselves up only to war and sloth. If a nest of Amazons (generally situated under a flat stone) be opened and disturbed, the masters run away without troubling themselves further, while the slaves with noble self-forgetfulness seize the pupæ and larvæ and try to save them. The Amazons are, therefore, in the true sense of the word filibusters and highwaymen, and direct their whole energy to robbery and slave-catching. Agreeably to this, the neuters are endowed with a personal courage which would seem to the highest degree marvellous to everyone if an ant were not in question, and which drives them to the most unheard-of exploits. A single Amazon, thrown into the middle of a number of hostile ants, does not seek to escape, as any other ant would do, but strikes right and left, piercing the heads of ten or fifteen opponents, until at last overpowered. This mad valor is only shown by the Amazon when it seems to know that it must be lost in any case; for when it fights in ranks and with the prospect of victory, its courage is of a wiser kind, and the single ant does not separate itself from the main army save under pressing need, and if necessary retreats with the latter. Only when the fight has become very embittered or has lasted for a long time, some Amazons at last grow so furious that, forgetting everything around them, they only take pleasure in blind biting and slaughter. They bite at everything they see, at pupæ, larvæ, and even pieces of wood. Forel saw them kill their own slaves which were trying to soothe them. Generally the slaves succeed in gradually quieting them down, and they are then able to find their way home, whereas during their fury they are incapable of doing this and run hither and thither as though they were mad. About twenty Amazons are generally enough to put to flight a crowd of enemies fifty times their number. Forel saw a small division of Amazons, numbering less than a hundred individuals, separate themselves from a great marauding expedition, and march against a very large nest of *F. rufibarbis*. Before reaching it they stopped still for a few moments as though frightened at their own boldness, and seemed to meditate. But they then flung themselves into the midst of their thousands of enemies, under whose numbers they soon disappeared. Several were seen to penetrate into the nest itself, in spite of the heavy hostile columns pressing out of it. Forel did not expect to see a single one come out again. The attacked ants appeared little disturbed, seeing the numerical weakness of their foes, and only here and there tried to save a pupa. The assailants, in spite of their audacity were unable to get much, and only about a third were seen returning laden with booty. Two or three Amazons were made prisoners, and the small number of the attackers was even useful to them, as the inhabitants of the nest rushed furiously about, without being able to seize them owing to their scattering. Their return was molested for a distance of two or three decimetres.

That which makes the Amazons so specially dreaded by the other ants is not their matchless courage so much as their manner of battle. They are not contented, like others of their race, with tearing off their enemies' legs, or feelers, or pieces of their bodies, or with biting the latter into pieces, and indeed the peculiar form of their mandibles, already described, renders this impossible; so they forthwith seize the head of their enemy and pierce it with their sharp strong jaws exactly at the place whereat the brain is situated. This attack generally succeeds and has immediate death as its result. But Forel one day saw a very large ant, belonging to the species *Atta structor*, wounded in this way, whose very hard head presented unexpected resistance. The attacking ant finally let its enemy loose, without being able completely to attain its object, and Forel noticed that the wounded ant was unable to move its jaws which had been mutilated by the bite, and were hanging loosely down. But it was able to run.

Most interesting naturally is the description of the marauding excursions and slave hunts themselves, undertaken from time to time by the Amazons, in order to bring to their nests the greatest possible number of pupæ of the slave species, which latter are destined to become regular slaves. These marauding excursions, and especially the wars and battles of the ants, which will be hereafter described, have so striking and surprising a resemblance to the wars and battles of men, that it seems as though ants had taken men, or men ants, for a model. For, to philosophers, standing above the strife of parties and contemplating men, the relative circumstances are equally ridiculous and contemptible, although the animal stands higher than the man in so far as it generally fights only for the sake of self-preservation, while among men the paltriest passions often give rise to his continual wars and quarrels, destroying the fruits of industry. Men will not emerge from their half-animal past and reach the full development for which they are destined, until everlasting peace and the universal brotherhood of nations shall have changed the present sad state of things for the attainment of the general good.

Lespès describes as follows a marauding excursion of the Amazons watched by himself: "These expeditions only take place towards the end of the summer and in autumn. At this time the winged members of the slave species (*F. fusca* and *F. cunicularia*) have left the nest, and the Amazons will not take the trouble to bring back useless consumers. When the sky is clear our robbers leave their town in the afternoon at about three or four o'clock. At first no order is perceptible in their movements, but when they are all gathered together they form a regular column, which then moves forward quickly, and each day in a different direction. They march closely pressed together, and the foremost always appear to be seeking for something on the ground. They are each moment overtaken by others so that the head of the column is continually growing. They are in fact seeking the traces of the ants which they propose to plunder, and it is scent that guides them. They snuff over the ground like hounds following the track of a wild animal, and when they have found it they plunge headlong forward, and the whole column rushes on behind. The smallest armies I saw consisted of several hundred individuals, but I have also seen some four times as large. They then form columns which may be five metres long, and as much as fifty centimetres wide. After a march, which often lasts a full hour, the column arrives at the nest of the slave species. The *F. cuniculariæ*, which are the strongest, offer keen opposition but without much result. The Amazons soon penetrate within the nest, to come out again a moment later, while the assailed ants at the same time rush out in masses. During the whole time attention is directed solely to the larvæ and pupæ, which the Amazons steal while the others try to save as many as possible. They know very well that the Amazons cannot climb, so they fly with their precious burdens to the surrounding bushes or plants, whereto their enemies cannot follow them. They then pursue the retreating robbers and try to take away from them as much of their booty as possible. But the latter do not trouble themselves much about them and hasten on home. On their return they do not follow the shortest road, but exactly the one by which they came, finding their way back by smell. Arrived at their nest, they immediately hand over their booty to the slaves, and trouble themselves no more about it. A few days afterwards the stolen pupæ or nymphæ emerge, without memory of their childhood, and immediately and without compulsion take part in all tasks."

They must gradually become accustomed to the war-expeditions and slave-hunts of their masters, for, according to Forel, they at first try to hold them back therefrom. They gradually come to regard them as a matter of course, and no longer oppose them, but give their masters a bad reception if they come back empty handed. Espinas ("Animal Societies," German ed., 1879, p. 362) confirms from his own experience the observation already made by Huber, " that Amazons which return empty-handed from an expedition are badly received and pulled about by the dark grey workers." Sometimes the slaves permit themselves freedoms and impertinences towards their masters which border on revolt and rebellion, but these are severely punished if they pass a certain point. For a time an Amazon will permit the importunity of its slave. But if it becomes too troublesome, it takes its head between its terrible jaws, and the slave submits forthwith. If it does not yield at once its death is certain. Forel one day saw an Amazon worried by six or seven slaves, pinched, irritated, pulled by the legs, etc. It put an end to the game by catching hold of one of them and piercing its head. A dry season often causes a rebellious disposition on the part of the slaves, the Amazons requiring drink from them too often, and the slaves being unable to meet such frequent demands. They become irritable and bad tempered and would seriously attack their masters if they were able to overcome them.

To come back to the marauding excursions. Lespès arrived at the conclusion that these expeditions were not carried out without ripe consideration and consultation, as well as preceding inquiries by special emissaries as to the hostile nests, which are often hard to find. Forel often saw single Amazons or small divisions leave the nests at different times during the day, and run about searching in various directions. These scouts serve later as leaders of the expedition. Forel also saw four or five inspecting a discovered nest of *F. fusca* and carefully investigating the entrances and the situation. This is the more necessary as the entrances are often difficult to find, so that in spite of all care and forethought the expeditions not seldom end without result. On the 29th July, 1873, in the afternoon at about five o'clock, Forel saw an enormous army of Amazons (about 1500) set out, and return home without finding anything. Another time he saw them vainly endeavor for a long time to penetrate into a nest of the *fusca*, the dome of which was quite closed, and communicated with the outer world by a subterranean canal opening at a distant spot. On another occasion it was a whole hour before they were able to find the entrance of a subterranean nest of the *fusca*. Lastly, Forel saw the leaders, before an expedition began, walk about on the surface of the nest for a long time, as though in consultation. Suddenly several of them went inside the nest, and soon afterwards out rushed crowds of warriors, which tapped each others' heads with their antennæ. Some, meanwhile, remained in the nest each time. The expedition was then arranged, the slaves paying no attention to the whole matter. The continual renewal of the head of the column arises from the fact, observed by

Lespès, that the original leaders remain behind to keep order in the rear, to guide it, and to inspire the laggards, while others take the place at the head. From time to time the army makes a short halt, partly to let the rearguard close up, partly because different opinions arise as to the direction of the host, or because the place at which they are is unknown to them. Forel several times saw the army completely lose its way—an incident only once observed by Huber. Forel puts the number of warriors in such an army at from one hundred to more than two thousand. Its speed is on an average a metre per minute, but varies much according to circumstances, and is naturally least when returning laden with booty. If the distance be very great, such bodily fatigue may at last be felt that the whole attack on the hostile nest is given up, and a retreat is begun; Forel once saw this happen after they had passed over a distance of two hundred and forty yards. Sometimes it seems as though, on coming within sight of the hostile nest, a kind of discouragement took possession of them, and prevented their making the attack. If the nest cannot at once be found, the whole army halts, and some divisions are sent forward to search for it, and these are gradually seen returning towards the centre. Forel also saw such an army only searching the first day, advancing zigzag, and with frequent halts, whereas on the following day it went forward to its aim swiftly and without delay, having found out the road. It seems that a single ant, even if it knows the way and the place, is not able alone to lead a large army, but that a considerable number must be employed in this duty. Mistakes as to the road occur with special ease during the return journey, because the several ants are laden with booty and cannot readily understand each other. Individual ants are then seen to wander about in every direction often for a long time, until they at last reach a spot known to them, and then advance swiftly to their goal. Many never come back at all. These mistakes easily occur when the robbers which have passed into a hostile nest do not come out again at the same holes whereby they entered, but by others at some distance, for instance by a subterranean canal. Coming out thus in a strange neighborhood they do not know which way to take, and only some chance on the right road during their aimless wanderings about, and recognise and follow it by smell. On the other hand such mistakes scarcely ever happen to individuals in an unladen train, kept in good array.

Other species of ants (*F. fusca, rufa, sanguinea*) know better how to manage under such circumstances, than do the Amazons. The laden ones lay down their loads, first find out where they are, and only take them up again after they have found their way.

If the booty seized in the nest first attacked is too large to be all taken away at once, the robbers return once or oftener so as to complete their work. The poor ravished folk try to stop up the entrances of their nest with earth as much as possible during the intervals, but the robbers open them again and go on with their plundering. When all is over, the attacked ants, whose resistance has only been feeble, bring back to the nest the larvæ and pupæ they have saved.

In carrying away the larvæ and pupæ the Amazons have to be very careful not to wound them with their pointed mandibles, whereas the other species easily accomplish this task with the help of their toothed jaws. Yet they sometimes forget this care in the heat and the excitement of the struggle, and kill their living burdens. They cannot, for the same reason, carry along very large pupæ, although their smaller and weaker slaves are able to perform this duty. Forel has admirably described (*loc. cit.* p. 225) their useless and droll attempts under such circumstances.

The ants, as already said, do not have regular leaders nor chiefs. Yet it is certain that in each expedition, alteration of road or other change, the decision during that event comes from a small knot of individuals, which have previously come to an understanding and carry the rest and the undecided along them. These do not always follow immediately, but only after they have received several taps on the head from the members of the "ring." The procession does not advance until the leaders have convinced themselves by their own eyesight that the main part of the army is following. How desirable an example for men, amongst whom it is generally said: *Tot capita tot sensus* (so many heads so many opinions), and amongst whom these divisions of opinions and interests so often mutilate or hinder the most important and useful undertakings.

One day Forel saw some Amazons on the surface of a nest of the *F. fusca* seeking and sounding in all directions, without being able to find the entrance. At last one of them found a very little hole, hardly as large as a pin's head, through which the robbers penetrated. But since owing to the smallness of the hole the invasion went on slowly, the search was continued, and an entrance was found further off, through which the Amazon army gradually disappeared. All was quiet. About five minutes later Forel saw a booty-laden column emerge from each hole. Not a single ant was without a load. The two columns united outside and retreated together.

A marauding excursion of the Amazons against the *F. rufibarbis*, a sub-species of the *F. fusca*, or small black ants, took place as follows: The vanguard of the robber army found that it had reached the neighborhood of the hostile nest more quickly than it had expected; for it halted suddenly and decidedly, and sent a number of messengers which brought up the main body and the rearguard with incredible speed. In less than thirty seconds the whole army had closed up, and hurled itself in a mass on the dome of the hostile nest. This was the more necessary as the *rufibarbes* during the short halt had discovered the approach of the enemy, and had utilised the time to cover the dome with defenders. An indescribable struggle followed, but the superior numbers of the Amazons overcame and they penetrated into the nest, while the defenders poured by thousands out of the same holes, with their larvæ and pupæ in their jaws, and escaped to the nearest plants and bushes, running over the heaps of their assailants. These looked on the matter as hopeless and began to retreat. But the *rufibarbes*, furious at their proceedings, pursued them, and endeavored to get away from them the few pupæ they had obtained, by trying to seize the Amazons' legs and to snatch away the pupæ. The Amazon lets its jaws slip slowly along the captive pupa as far as the head of its opponent and pierces it, if it does not, as generally happens, draw back. But it often manages to seize the pupa at the instant at which the Amazon lets it go and flies with it. This is managed yet more easily when a comrade holds the robber by the legs, and compels it to loose its prey in order to guard itself against its assailant. Sometimes the robbers seize empty cocoons and carry them away, but they leave them on the road when they have discovered their mistake. In the above case the strength of the *rufibarbes* proved at last so great that the rearguard of the retreating army was seriously pressed, and was obliged to give up its booty. A number of the Amazons also were overpowered and killed, but not without the *rufibarbes* also losing many people. None the less did some individuals, as though desperate, rush into the thickest hosts of the enemy penetrated again into the nest and carried off several pupæ by sheer audacity and skill. Most of them left their prey to go to the help of their comrades when assailed by the *rufibarbes*. Ten minutes after the commencement of the retreat all the Amazons had left the nest, and, being swifter than their opponents, they were only pursued for about half way back. Their attack had failed on account of a short delay!

On another occasion observed by Forel, in which several fertile Amazons also took part and killed many enemies, the nest was thoroughly ravished, but the retreat was also in this case very much disturbed and harassed by the superior numbers of the enemy. There were many slain on both sides. That in spite of the above-mentioned unanimity different opinions among the members of an expedition sometimes hinder its conduct, the following observation seems to shew: An advancing column divided after it had gone about ten yards from the nest. Half turned back, while the other half went on, but after some time hesitated and also turned back. Arrived at home it found those which had formerly turned back putting themselves in motion in a new direction. The newly returned followed them, and the re-united army, after various wheelings, halts, etc., at last turned home again by a long way round. The whole business looked like a promenade. But apparently different parties had different nests in view, while others were entirely against the expedition. Yet perhaps it was only a march for exercise.

Outer obstacles do not, as a rule, hinder the Amazons when they are once on the march. Forel saw them wade through some shallow water, although many were drowned in it, and then march over a dusty high road, although the wind blew half of them away. As they returned, booty-laden, neither wind, nor dust, nor water could make them lay down their prey. They only got back with great trouble, and turned back again to bring fresh booty although many lost their lives.

The scenes which occur during the marauding excursions are as manifold and as various as during the wars and marauding expeditions of men, and like these might either be sung in the style of the epic poem or written of as in a general's dispatches, if they did not concern ants. An Amazon column goes back to a nest that it has already partially ravished, in order to finish its work, but has become rather scattered on the way. The small number of robbers marching in front are seized and taken prisoners by the assailing *rufibarbes*, which have meanwhile assembled. The Amazons following, seeing this, halt and await the main army. After its arrival the *rufibarbes* are attacked and flung aside in heaps. The prisoners are freed, and fresh pupæ carried away.

The *rufibarbes* preserve their warlike character even as slaves. If hostile ants approach the nest in which they live as slaves, they gallantly attack them, while the small black slaves (*F. fusca*) are content to call their masters, and only rarely take part themselves in the fight. On the other hand, they help diligently in the plundering, when their owners have destroyed a hostile nest or a hostile army. Forel saw them pick up and carry back to the nest some of their masters which had lost their way in the heat of battle.

The most terrible enemy of the Amazons is the sanguine ant (*F. sanguinea*), which also keeps slaves and thereby often comes into collision with the Amazons on their marauding excursions. It is not equal to it in bodily strength or fighting capacity, but surpasses it in intelligence ; according to Forel it is the most intelligent of all the species of ants. If Forel, for instance, poured out the contents of a sack filled with a nest of the slave species near an Amazon nest, the Amazons apparently generally regarded the tumbled together heap of ants, larvæ, pupæ, earth, building materials, etc., as the dome of a hostile nest, and took all imaginable but useless pains to find out the entrances thereinto, leaving on one side for this investigation their only object, the carrying off the pupæ : but the sanguine ants under similar circumstances did not allow themselves to be deceived, but at once ransacked the whole heap. On August 3, 1869, Forel set down an apparatus containing an artificial nest of Amazons with small black slaves (*F. fusca*) quite close to a nest of the

L

sanguine ants. The slaves came out first, and were at once seized by the sanguine ants and slain in great numbers. But as these approached close to the Amazon nest, about a dozen Amazons rushed out and flung themselves on the intruders. Their small number, however, disappeared in the masses of their enemies, which manifested an even more menacing disposition. Now came new warriors out of the nest to help, and although far fewer in number than the sanguine ants, flung the assailants back to the entrance of their own nest. Among them, Forel saw a single ant engaged in combat with ten or twenty opponents. But this was not enough ; the Amazons followed their enemies, which seemed overcome by panic, right into their nest and chased out all the inmates, and then began such a plundering, and such numbers of pupæ were brought out, that the entrance of the Amazon nest began to be stopped up. About thirty Amazons had fallen, most of them by the poison of their enemies, while heaps of dead sanguine ants covered the ground. The ravished pupæ or nymphæ were eaten or thrown away by the slaves if they belonged to the sanguine species, while those belonging to slave species were carried into the nest to be brought up as slaves. Only after many days did the driven away sanguine ants, which had meanwhile taken refuge in the neighboring grass, venture back to their nest.

Another time Forel saw the two slave-making species meet during a marauding excursion. An Amazon army was marching in rather loose order, when Forel noticed at a distance of only a few decimetres a number of *rufibarbes* which had fled to bushes and stalks with pupæ in their mouths. This led him to the discovery of a *rufibarbis* nest, which had just been plundered by the sanguine ants, on the dome of which a large number of the robbers were still promenading. Meanwhile the Amazon army had come up, and flung the sanguine ants, which fled in every direction, off the ant heap, but could find nothing left in the empty nest. They thereupon angrily followed the sanguine ants, which scattered and hid in the grass, no great results following. One day (August 12) the inhabitants of an Amazon nest were warming themselves in the sun, hanging motionless and in masses from the grass-stalks sprouting over their nest. A signal was suddenly given which set them all in movement. Only some which were too far off appeared not to have understood it and remained quiet. But they also ran rapidly up when they saw the general commotion. The procession moved off towards a nest of the *fusca*, but before it arrived Forel had poured out on it a sackful of sanguine ants kept ready, and had made a breach in the nest. The sanguine ants pressed in, while the *fuscæ* came out to defend themselves. At this moment the first Amazons arrived. When they saw the sanguine ants they drew back and awaited the main army, which appeared much disturbed at the news. But once united, the bold robbers rushed at their foes. The latter gathered together and beat back the first attack, but the Amazons closed up their ranks and made a second assault which carried them on to the dome and into the midst of the enemy. These were overthrown, as well as a number of *F. pratensis*, which Forel at this moment poured out on the nest. The conquerors delayed for a moment on the dome after their victory, and then entered the nest to bring out a little of the valuable booty. A few Amazons which were mad with anger did not return with the main army, but went on slaughtering blindly among the conquered and the fugitives of the three species, *fusca*, *pratensis*, and *sanguinea*.

The ravished *rufibarbes* once became so desperate at their overthrow that they followed the robbers to their own nest, and the latter had some trouble in defending it. The *rufibarbes* let themselves be killed by hundreds, and really seemed as though they courted death. A small number of the Amazons also sank under the bites of their enemies. The nest contained slaves of the *rufibarbis* species, which on this emergency fought actively against their own race. There were also slaves of the species *fusca*, so that the nest included three different species of ants.

The same nest is often revisited many times on the same day or at different periods until either there is no more to steal, or the plundered folk have hit upon better mode of defence. A column which was in the act of going back to such a plundered nest turned when half-way there, and halted, apparently on no other ground than because it had met the rearguard of the army, and had learned that the nest was exhausted, and that there was nothing more to be had there (Forel, *loc. cit.*, p. 318). The robbers then went off to a *rufibarbis* nest which was in the neighborhood, and killed half the inhabitants while plundering the nest. The surviving *rufibarbes* returned after the robbery and brought up new progeny; but thirteen days later the Amazons again reaped a rich harvest from the same nest. The Amazon army often severs itself into two separate divisions when there is not enough for both to do at the same spot. Sometimes one division finds something and the other nothing, and they then reunite. If any obstacle be placed in their way they try to overcome it, in doing which some leave the main army, lose themselves, and only find their way home again with difficulty. Forel has tried to establish the normal frequency of expeditions, and found that a colony watched by himself for a space of thirty days sent out no less than forty-four marauding excursions. Of these about eight and twenty were completely, nine partially, and the remainder not at all successful. He four times saw the army divide into two. Half the expeditions were levelled against the *rufibarbes*, half against the *fuscæ*. On an average a successful expedition would bring back to the colony a thousand pupæ or larvæ. On the whole, the number of future slaves stolen by a strong colony during a favorable summer may be reckoned at forty thousand!

The internecine battles which occasionally break out among the Amazons themselves are naturally the most cruel. They tear each other to pieces with incredible fury, and knots of five or six individuals which have pierced each other may be seen rolling over each other on the ground, it being impossible to distinguish between friend and foe. Civil wars among men also are known to be the most embittered and the most bloody.

CHAPTER XII.

SLAVERY — *(Continued).*

THE habits of the *F. sanguinea,* or sanguine ant, the second European slave-making species, much resemble those of the Amazons. But there is the essential difference between them that the sanguine ants are not, like the Amazons, wholly dependent upon their slaves, for they are able to work and can feed themselves. They therefore keep their slaves more as helpers than as servants, and do not require nearly so large a number of them as do the Amazons. They can even do without them, for Forel has often found a nest of the *sanguinea* quite without slaves, as in the Maloggia Pass at the foot of Mount Tendre and elsewhere. The marauding excursions of the *sanguinea* are also far less frequent than those of the Amazons and only take place in each colony two or three times a year. The *fusca* and the *rufibarbis,* or *cunicularia,* here also generally provide the slaves, which work in common with their masters at building dwellings and roads, as well as attending the larvæ, pupæ, etc.; occasionally slaves are taken from other species.

The sanguine ants are generally fond of honey, and tear to pieces living insects in order to lick up the juices of their bodies. They do not spare the slave-pupæ brought in, and sometimes even eat their own eggs, larvæ, and nymphæ, as well as those of other species which they do not generally enslave. They are quite able to distinguish the nymphæ of the males and fertile females of their slaves from those of the workers, and kill the former while they spare the latter. When Forel offered them a living wasp, it was seized by four workers, sprinkled with poison and strangled. The corpse was then torn in pieces. Amazons, which Forel had brought up artificially with sanguine ants and four or five other species, showed none of their usual ferocity and were quite peaceable, and all those brought up together lived happily with each other, giving each other honey, etc. Even when Forel set them at liberty they remained together and all betook themselves to their new dwelling. And then occurred the following noteworthy episode. A little sanguine ant wanted to take hold of an Amazon and carry it off. The carried ant, during the transit, generally rolls itself round the head of its bearer, so as to lighten its weight as much as possible. But as the Amazon did not or would not perform this duty, the sanguine ant contented itself with seizing one leg and marching off towards the new nest. As the Amazon resisted, but did not try to bite, the affair proceeded slowly. After a while the sanguine ant let its comrade go in order to reconnoitre. Meanwhile the Amazon ran uneasily hither and thither. A chance passer-by, a *rufa*, or hill ant, belonging to the community saw this, and tried on its part to carry the Amazon on. But the latter still resisted, and as the hill ants are rather clumsy it failed in the attempt. The sanguine ant now came back and touched the *rufa* several times with its feelers. The communication must have been satisfactory, for the latter let go and resigned the Amazon to its original bearer, which now pulled it along to the new nest.

This artificial friendship between distinct and usually hostile species plainly shows how much education and the first impressions of childhood and youth may change and influence even inherited character and the pretended "instinct."

In their method of fighting, the *sanguineæ* show special refinement and circumspection. When they have to do with an enemy equal to them in strength they never attack it in front, but always try to assail its flanks. They generally march in small divisions, which continually send out scouts and spies, partly to bring up the rear and partly to find out the movements and the weak side of the enemy. When they have to do with a regular army of the large *F. pratensis*, they try to terrify it by surprises. They send several divisions to the flanks and rear of the foe, and then rush with incredible impetuosity into the midst of the hostile army, but retreat the moment they find that the resistance is too strong. They know how to unite courage and prudence, and understand, just like men, how to "fall back to the rear" when it is necessary, as military phraseology has it. They, however, generally succeed in their object, as the *pratenses* are struck with fear by this attack from behind, and give way. At such moments is shown the intelligence of the *sanguineæ*. They know how to seize the exact moment at which the enemy shows signs of retreat, and spread the intelligence from one to another with marvellous rapidity. They unhesitatingly throw themselves amongst the foe, slay right and left like the Amazons, and tear away the pupæ from their opponents. The *pratenses* are in such a state of consternation that they would not offer the least resistance if they were hundreds against one. Forel saw some conquered *pratenses* which had fled in crowds with their pupæ beneath the broad leaves of a plantain. A single sanguine ant pushed in among them, whereupon they all fled leaving their pupæ behind. No species of ant has such a desire for alien pupæ as has the *sanguinea;* to steal them seems the one object of its life. While the *pratensis*, for example, gets angry with its opponents and murders its prisoners, the *sanguinea* scarcely ever does anything of the kind. It conquers its enemies less by killing them than by spreading terror in the hostile camp. Forel has often seen the same warrior rob its enemies of pupa after pupa without molesting them further, and without being able to carry away all the stolen pupæ. Does it intend merely to spread fear, or to prevent the enemy from carrying away the pupæ? It is certain that it secures both ends.

" In the genus *Formica*, which I regard as the most intelligent of ant genera," says Forel (*loc. cit.*, p. 443), "the *sanguinea* without doubt carries off the palm. No other species shows so much diversity in its habits and in its ability to adapt itself to circumstances. It takes its slaves from a number of other species, fights with wonderful tactical skill, builds its nest in every imaginable way according to the place at which it chances to find itself, invents different kinds of attack for different enemies (*L. niger, F. pratensis, F. fusca*)" and so on.

As to the slave-hunts of the *sanguinea* they have already been described by Huber in unsurpassable fashion. Since then Charles Darwin has also taken the trouble to watch the slave-habits of this species, whereon we had better let the celebrated naturalist speak for himself:—

" I opened fourteen nests of *F. sanguinea*, and found a

few slaves in all. Males and fertile females of the slave species (*F. fusca*) are found only in their own proper communities, and have never been observed in the nests of *F. sanguinea*. The slaves are black and not above half the size of their red masters, so that the contrast in their appearance is very great. When the nest is slightly disturbed, the slaves occasionally come out, and like their masters are much agitated and defend the nest: when the nest is much disturbed and the larvæ and pupæ are exposed, the slaves work energetically with their masters in carrying them away to a place of safety. Hence, it is clear, that the slaves feel quite at home. . . . . .

"One day I fortunately witnessed a migration of *F. sanguinea* from one nest to another, and it was a most interesting spectacle to behold the masters carefully carrying (instead of being carried by, as in the case of *F. rufescens*) their slaves in their jaws. [Not always. L. B.] Another day my attention was struck by about a score of the slave makers haunting the same spot, and evidently not in search of food; they approached and were vigorously repulsed by an independent community of the slave species (*F. fusca*); sometimes as many as three of these ants clinging to the legs of the slave-making *F. sanguinea*. The latter ruthlessly killed their small opponents, and carried their dead bodies as food to their nest, twenty-nine yards distant; but they were prevented from getting any pupæ to rear as slaves. I then dug up a small parcel of the pupæ of *F. fusca* from another nest, and put them down on a bare spot near the place of combat; they were eagerly seized, and carried off by the tyrants, who perhaps fancied that, after all, they had been victorious in their late combat.

"At the same time I laid on the same place a small parcel of the pupæ of another species, *F. flava*, with a few of these little yellow ants still clinging to the fragments of the nest. This species is sometimes, though rarely, made into slaves, as has been described by Mr. Smith. Although so small a species, it is very courageous, and I have seen it ferociously attack other ants. In one instance I found to my surprise an independent community of *F. flava* under a stone beneath a nest of the slave-making *F. sanguinea;* and when I had accidentally disturbed both nests, the little ants attacked their big neighbors with surprising courage. Now I was curious to ascertain whether *F. sanguinea* could distinguish the pupæ of *F. fusca*, which they habitually make into slaves, from those of the little and furious *F. flava*, which they rarely capture, and it was evident that they did at once distinguish them: for we have seen that they eagerly and instantly seized the pupæ of *F. fusca*, whereas they were much terrified when they came across the pupæ, or even the earth from the nest of *F. flava*, and quickly ran away; but in about a quarter of an hour, shortly after all the little yellow ants had crawled away, they took heart and carried off the pupæ.

"One evening I visited another community of *F. sanguinea*, and found a number of these ants returning home and entering their nests, carrying the dead bodies of *F. fusca* (showing that it was not a migration) and numerous pupæ. I traced a long file of ants burthened with booty, for about forty yards, to a very thick clump of heath, whence I saw the last individual of *F. sanguinea* emerge, carrying a pupa; but I was not able to find the desolated nest in the thick heath. The nest, however, must have been close at hand, for two or three individuals of *F. fusca* were rushing about in the greatest agitation, and one was perched motionless with its own pupa in its mouth on the top of a spray of heath, an image of despair over its ravaged home."

So far Darwin. But while he only gives a general picture Forel here also relates many detailed observations of the most interesting kind. He observed the expeditions of the *sanguinea* to steal pupæ in Switzerland (Waadtland) from the middle of June to the middle of August. They march in small troops which, in case of need, summon reinforcements and therefore, as a rule, only reach their goal slowly. Between the individual troops messengers or scouts run continually backwards and forwards. The first troop which arrives at the hostile nest does not rush at it, as do the Amazons, but contents itself with making provisional reconnaissances, wherein some of the assailants are generally made prisoners by the enemy, which have time to bethink and to collect themselves. Reinforcements are now brought up and a regular siege of the nest begins. A sudden invasion, like that of the Amazons, is never seen. The besieging army forms a complete ring round the hostile nest, and the besiegers hold this with mandibles open and antennæ drawn back, without going nearer. In this position they beat off all assaults of the besieged, until they feel themselves strong enough to advance to the attack. This attack scarcely ever fails, and has for its chief object the mastering of the entrances and outlets of the nest. A special troop guards each opening, and only allows such of the besieged to pass out as carry no pupæ. This manœuvre gives rise to a number of comical and characteristic scenes. By this means the sanguine ants in a few minutes manage to have all the defenders out of the nests and the pupæ left behind. This is the case at least with the *rufibarbes*, while the rather less timid *fuscæ* try, even at the last moment when it is useless, to stop up or barricade the entrances. The sanguine ants do not indeed possess the terrible weapons and the warlike impetuosity of the Amazons, but they are stronger and larger. If a *fusca* or a *rufibarbis* fights with a sanguine ant for the possession of a pupa, it is generally very soon overcome. While the main part of the army is penetrating into the nest to steal the pupæ, some divisions pursue the fugitives, to take away from them the few pupæ which may chance to have been saved. They drive them even out of the cricket-holes in which they have meanwhile taken refuge. In short it is a *razzia*, or sweeping burglary, as complete as can be imagined. In the retreat the robbers in nowise hurry themselves, for they know that they are threatened by no danger and no loss, and the complete emptying of a large and distant nest often takes several days in accomplishing. The ants which have been so thoroughly robbed scarcely ever return to their former abode.

It must be admitted that a human army, robbing a foreign town or fortress, could not behave better, more prudently, nor with more regard to circumstances than do these wonderful animals.

As the only object of these expeditions is robbery, the thieves do not, as a rule, trouble themselves about the murder of their enemies, if these offer no active resistance. It is only when the latter hang on their legs and will not let go that they tear them in pieces with their jaws, for nothing is more unpleasant to them than being held by the leg. Their moderation, however, fails when they are merely conquering a strange nest, or are stealing such pupæ as they use for food, as of the *Lasius niger* or *flavus*; in these cases they

kill the inhabitants without mercy, and take pleasure in either changing into or living at the same time in such nests near their own. They are also not content to have one house or castle, but are like princes and rich people, who have several houses, castles, or villas at their disposal, while poor people have none at all or live in bad ones, like the horses and dogs of the rich. Forel knew one colony of the sanguine ants which had three nests and lived in them in turn!

When the sanguine ants are conquered, as sometimes happens, as by the strong *pratenses*, or when their foes are in too great numerical superiority, they understand how to retreat in good order and defend to the uttermost the entrances of their nest. The *pratenses* generally manage the blockading of the nest so clumsily, that the sanguine ants have time to fly with their pupæ by the further outlets. The only complete overthrow suffered by the sanguine ants is met with, as already related, at the hands of the Amazons, which unite the same tactics to better weapons, greater decision, and more vigorous massed assaults. The Amazons also march more rapidly and understand signals better. They conquer the sanguine even more easily and quickly than other species, because there is a certain amount of anxiety in conjunction with their prudence, and they are more easily frightened by a sudden attack. Above all it appears as if the impetuous assaults and surprises among most war-making ants had for their chief object the spreading of sudden panic in the hostile camp. Every kind of stratagem is also practised if it can serve their object. Forel saw a laden army of Amazons returning from a marauding expedition suddenly attacked by a small troop of sanguine ants. A part of the Amazons laid down their pupæ in order to fight better. The sanguine ants took advantage of this moment to seize the deposited pupæ and ran off with them.

After the Amazons the most powerful foe of the sanguine ants is the often mentioned *pratensis*. Yet Forel managed, by artificial mixing, to get the latter to completely serve as slaves in some nests, and both lived together on the best terms. If the *pratenses* are very numerous, the architecture of the nest takes quite the character of their peculiar style of building. Only the latter are, as a rule, seen promenading on the dome of the nest. But at any alarm or on the approach of hostile troops most of them fly into the interior to seek help; and in a moment the dome is reddened by an eruption of sanguine ants. If a stranger *pratensis* be thrown on the nest, those at home will fight against their own brother with similar, or even greater, anger than do their masters. During a change of abode Forel saw the sanguine ants seize and carry off the *pratenses*, while some of the latter turned back again. It seems as though attachment to an old dwelling were stronger in the latter than in the former, and this may arise from the *pratenses* being better architects.

The third slave-holding species in Europe is the *Strongylognathus*, a small *Myrmica* species, which takes its slaves from the *Tetramorium cæspitum*, or turf ants, also belonging to the genus *Myrmica*. It is rather rare and much resembles the Amazons in its habits, and like them has long pointed mandibles which make it incapable of work. On the other hand it can feed itself, although it does so unwillingly and prefers to be fed by its slaves. This is done, as Lespès relates, in a very peculiar manner, for its own jaws and the rather long jaws of its slaves make the usual *rapprochement* of the two mouths very difficult if not impossible. It seizes the slave, lays it gently on its back, and lets itself be fed in this position, in which the two mouths match. It also lets itself be carried by its slaves, which are ten times as numerous as the masters. The marauding excursions appear to take place at night, for the *Strongylognathus* have never been seen marching out during the day.

Forel, who distinguishes two sub-species, *S. testaceus* and *S. Huberi*, calls the first do-nothings or sluggards, and a pitiful caricature of the Amazons. Like them it tries to kill, but is generally too feeble to succeed. The defence of the nest is carried out more by the slaves than by the masters. Yet the latter are brave and, like the Amazons, fling themselves into the midst of the enemy, biting quickly in all directions, whereat the foes are more alarmed than they need be, for they seldom succeed in killing an opponent while they lose their own lives. They are even too weak to carry away their enemies' pupæ, and make the most comical efforts over them, whereas their slaves manage the task very easily. Without the latter, *S. testaceus* would be quite unable to plunder a hostile nest. Apparently the whole species is retrograde or undergoing retrogression, a condition easily comprehensible on the principles of evolution.

Forel bestows a somewhat better character on the *S. Iluberi*. He saw an army of these overthrow a colony of the *Tetramorium* with skill and courage, without the help of their slaves. In this case, as with the Amazons, their pointed mandibles spread great terror among their opponents, although they were rarely able to pierce their heads. Many of them remained on the field of battle, the *Tetramoria* seizing them on the breast with their jaws, and dismembering them. This species also does no work, and lets itself be fed by its slaves; only in case of need does it condescend to feed itself.

Before we quit the interesting question of slavery let us throw a rapid glance on the slave species themselves, or at least on the chief of them.

The most remarkable of these is the *F. rufibarbis* or *cunicularia*, a sub-species of the *F. fusca* or small black ant, the species which yields most of the slaves. It manifests remarkable bravery and skill in its battles, whether with slave hunters or with other species. It is able to chase the rather clumsy *F. rufa*, or hill ant, away from its pupæ, even when the enemy is superior in numbers. If a *rufa* catch hold of its leg, it either kills or disables it, and if the *rufa* let go, it runs away. A badly guarded pupa is at once seen and snatched away, and if a *rufa* lets go its pupa for an instant, so as to get a better hold, it is at once carried off. Let a single *rufibarbis*, says Forel, be put in the midst of *rufæ* or *pratenses*, covering the dome of a nest, and it will almost always be found to escape scot free. It also catches flies at its breeding time. On the other hand it does not understand, as do other species, how to introduce order and tactics into its united movements, whereas the individually less skilful *rufæ* always fight in close columns and sacrifice themselves without hesitation to the common good. A small troop never divides off to make a side attack, and no single ant goes out by itself as an adventurer. It is also unable to pursue a flying foe.

Perty (*loc. cit.*, p. 334) says that he once, at Berne, beat with his umbrella a hazel-nut bush on which countless *F. rufæ* were busy, and then picked up a few of them. A specially large ant, "which by its whole behavior seemed as though it recognised me as the disturber" rose threateningly half erect, and bit him in the finger.

C. Shröder, of Elberfeld ("Zool. Gardens," 1867, p. 225) tried to give a *rufa* as food to a so-called Aphis-lion*, a strong robber insect larva. But the brave ant turned the tables and so faced its terrible foe that the latter stayed where it was. The same observer had the opportunity of watching the remarkable behavior of an ant-slave of the *F. fusca* imprisoned with some red-brown ants and three myrmecophilous insects (*Lomechusa strumosa*). It alone did more than all the rest put together, carried the larvæ and pupæ to a place of safety, looked after the nursing and feeding, etc., and seldom left the nursery.

The often-named *F. pratensis* is a sub-species of *F. rufa*, and has colonies or nests containing from 5,000 to 500,000 individuals. Forel observed among them a remarkable case of nursing the sick on the occasion of a change of abode. On the dome of the old nest an ant, which was clearly ill, was walking with tottering steps, drooping feelers, and half-closed jaws. Other ants approached it, licked it, and contemplated it from various points of view, trying gently to draw it within the nest. Suddenly up came one of the out-goers, pushed the others aside and tried to lay hold of the sick ant. It invited it to take hold of one of its mandibles, but the invalid did not seem to understand. After prolonged and futile trouble, the latter at last folded back its legs and antennæ, and let itself be taken up by its companion, which carried it to the new nest. A quarter of an hour later Forel saw the pair still on their road, recognising them by the unusual way in which the sick ant was carried. He pulled them apart with a straw, and the sick one hobbled on. But its friend came back as soon as it had recovered from its fright and once more carried off the invalid.

A still more striking instance of care of the sick was observed by Moggridge (*loc. cit.* p. 46), who saw an ant (*Atta*) pull a sick comrade to a little pool, bathe it for

---

* [In Adelung's dictionary it is stated that the Aphis-eater, which is probably the same as the Aphis-lion, lives on the leaves of elms, changes into a fly, and lays in wait for the Aphides. It was first minutely described in 1770, and is ranked by Linnæus among the *Hemerobii*. Tr.]

several seconds, and then bring it back with great trouble to dry it in the sun, and perhaps to let it recover its strength. These cases are the more surprising, as ants usually are wont to desert those which are very sick, or to throw them out of the nest, as is seen by Ebrard's experiments.

It was of the *pratensis* that Huber wrote the observations touching its gymnastic sports which became so famous. He saw these ants on a fine day assembled on the surface of their nest and behaving in a way that he could only explain as simulating festival sports or other games. They raised themselves on their hind legs, embraced each other with their forelegs, seized each other by the antennæ, feet, or mandibles and wrestled—but all in friendliest fashion. They then let go, ran after each other, and played hide and seek. When one was victorious, it seized all the others in the ring, and tumbled them over like ninepins.

This account of Huber's found its way into many popular books, but in spite of its clearness won little credence from the reading public. "I found it hard to believe Huber's observation," writes Forel, "in spite of its exactness, until I myself had seen the same." A colony of the *pratensis* several times gave him the opportunity, when he approached it carefully. The players caught each other by the feet or jaws, rolled over each other on the ground like boys playing, pulled each other inside the entrances of their nest only to come out again and so on. All this was done without bad temper, or any spirting of poison, and it was clear that all the rivalry was friendly. The least breath from the side of the observer was enough to put an end to the games. "I understand," continues Forel, "that the affair must seem marvellous to those who have not seen it, especially when we remember that sexual attraction can here play no part."

CHAPTER XIII.

FRIENDSHIPS AND ENMITIES.

THE capability of friendship, both social and individual, is as much developed among ants as is that of war and enmity. Also quite apart from the slave hunts, already described, and undertaken for the attainment of a special object, war and battle are the universal watchword of almost all ant species, so that it may fairly be said that the ant is the ant's worst enemy. Only some small and peaceful species, such as the *Botryomyrmex meridionalis*, form an exception. As to the object of the continual and usually very bloody wars, they seem often to be carried on just as among men, without any decided ground and from mere lust of fighting and slaying. At other times they are struggles for property and land, for soil and subsoil, plant lice, etc.; at other times again they are robberies of pupæ which serve for food, or of stored up provisions, as among the harvesting ants, or the conquest of a new dwelling. Generally it may be said without exaggeration that all the dwellers in the same nest or the same colony are friends, all inhabitants of strange or different nests or colonies are enemies; and if different species generally fight most frequently and eagerly, yet often the most embittered battles occur between different nests of the same species. If two hostile ants meet each other, they generally fly from or avoid each other, but if one of them is much larger and stronger or knows itself to be supported by friends it will then attack the other. If two friendly ants meet they either run silently past each other, or else hold a conversation by means of their antennæ. Such a tapping of feelers also takes place when each is in doubt whether it has to do with an enemy or a friend. If the latter be the case and the friend finds that the abdomen of the other is well filled, it begs, coaxing and stroking with its feelers, for a little refreshment by the road, which is willingly given in the way already described. According to Forel, mutual feeding is a sure sign of friendship. The same is true of mutual carrying, which is sometimes practised, in order to show the friend a new road or place, sometimes to obtain help for a given work at a certain spot. On a change of dwelling those which know the way carry those which do not know it, and workers which are weary with a long journey are also carried by their companions. It sometimes happens that a conquered enemy is seized by the victor and carried or pulled as a prisoner into the nest, but this can be recognised by the particular way of carrying. There is also very different behavior in the two cases if the pair are forcibly separated, in the one friendly recognition on meeting again, in the other flight or renewed strife. Also the carrying of enemies is rare, while that of friends is a very frequent occurrence.

If friendly ants are put together in a box they quickly recognise each other, and carry, feed, and lick each other; if they are hostile they fall on or avoid one another.

W. M. R. ("Nature," 1879, No. 4) put some little flasks covered with muslin, some containing friendly or related, and the others hostile or stranger ants, before the nests of various species (*F. fusca, P. rufescens*, etc.). The former were each time left quite undisturbed and unnoticed, while the latter were immediately attacked. They tried to bite through the muslin, and as they could not succeed in doing this they set sentinels in front of the hostile camp. At last the muslin was bitten through, and the imprisoned ants would soon have been killed if the observer had not taken them away. W. M. R. considered that this experiment proved that the feeling of hatred was stronger among these remarkable creatures than that of love. An exactly similar observation was made by Sir John Lubbock (*loc. cit.*), and he found the animals he was experimenting on bit through the muslin and killed those inside. If one of the latter stretched a leg outside it was at once laid hold of, and the enemy tried to pull it through. The same observer came to the conclusion that his experiments proved that individual friendships were to be found among ants. Some have very many, others few, others again, as it seems, no friends. Specimens of *F. fusca*, for example, never brought friends to a foodstore they had discovered, while all others did.

If two hostile ants meet, each of which knows that it can reckon on the help of its comrades, a murderous conflict generally takes place, in which mandibles, stings (when they have them) and poison play the chief part. They also snatch at each others' legs and try to pull each other into the hostile camp, wherein swift justice is dealt out to the vanquished. The battle is decided most quickly when one succeeds in seizing the thorax of its rival, and either pulls or bites off the head or at least destroys the large nervous cord that runs down the middle line of the body; each therefore tries to the utmost to guard itself against this manœuvre, so that it only succeeds either by surprise or when one of the combatants is much larger than the other. Among ants with bad eye-sight battles are much slower affairs than among those with good, as they guide themselves almost only by their feelers. Sometimes the combatants or the victors manifest a really infernal cruelty, which almost approaches the wickedness shewn by man to man. They slowly pull from their victim, that is rendered defenceless by wounds, exhaustion, or terror, first one feeler and then the other, then the legs one after another, until they at last kill it, or pull it in a completely mutilated and helpless condition to some out-of-the-way spot where it perishes miserably. Yet some compassionate hearts are to be found among the victors, which only pull the conquered to a distant place in order to get rid of them, and there let them go without injuring them.

If a single ant be seized at the same time by several enemies it is, as a rule, lost. For while it is held fast on every side, one of its opponents springs on its neck and tries to bite through it. Sometimes it is only taken prisoner and dragged into the hostile nest, there to be slaughtered in the cruelest manner. A vanquished ant in dying often clasps itself so firmly round the limbs of its enemy that the latter has the greatest trouble to get free. Often its comrades can only get the corpse away in pieces, while the head is not seldom carried about for days, until decay makes it fall off. Only the Amazons are able to avoid this misfortune by piercing through the head of their enemies in the way already described, for the jaws of their enemy lose their strength when the brain is destroyed.

Among most ants their courage rises in proportion to the

number of their comrades or the size of their colony. The same ant which shuns no danger when joined by many others, becomes anxious and timid when it knows itself to be alone or surrounded by few companions. Perhaps, also the tendency to self-preservation and care for the safety of smaller colonies or societies lead to the avoidance of serious danger and conflicts, while larger societies see no harm in sacrificing some of their citizens.

The wounded and the sick, as already mentioned, are taken care of. If their case is regarded as hopeless, they are carried to a distant place and left there to die. In similar fashion, at the end of a fight, the corpses, or the remains of them, are taken out of the nest, for the ants keep their nests as clean as they do their bodies. Dupont maintains that many species of ants have their own church-yards or common burial-grounds, and that they formally deposit therein their dead or their fallen. Improbable as this may sound, several casual observations have been made which prevent us from relegating it quite to the region of fable. After a battle artificially brought about in a garden between four different species (*rufa, sanguinea, cinerea,* and *pratensis*), Forel saw the field of battle covered with the slain of each. But the greater number were, remarkable to say, arranged in a long and regular row as though they were to be buried. Perty (*loc. cit.* p. 318) publishes a communication from Mrs. Lewis Hatton, of Sidney, which relates a regular burial of twenty ants pressed to death by their companions. Bingley (*loc. cit.* p. 174) also mentions the observation of an Englishman who saw an ant bring the body of a comrade out of the nest and carry it away to a distant place. This he saw repeated several times, one after another. Lubbock (*loc. cit.*) also saw dead ants carried out of the nest and laid in heaps at a distant spot "exactly as in a cemetery." McCook (*loc. cit.*) saw similar proceedings among all the species which he observed. The corpses were always heaped together in an out-of-the-way place, "as though the crude idea of a mortuary had dawned on the mind of the little creatures."

In all these cases the dead of their own friends have been in question, but the corpses of enemies are generally torn in pieces in order to lick up the sweet juices therein contained. On the other hand, as Forel assures us, the grown ants of the same nest never seize each other with this intent, although many species, as already mentioned, eat their own larvæ and pupæ. They would rather die of hunger. Cannibals among men are not so tender-hearted.

It is wonderful also to see the way in which friendly ants —and those which have been friendly—recognise each other after long separation, and distinguish between friends and foes, however many of them there may be, or even when they belong to the same species. Darwin has given much attention to this characteristic and has spoken of it in several of his works. He several times brought ants of the same species (*F. rufa*) from one ant hill to another, which, as it seemed, was inhabited by tens of thousands of ants, yet the strangers were at once recognised and killed. Even the sprinkling of ants with assafœtida does not prevent them from being recognised by their comrades, so that not smell, but an unknown something, perhaps a sign or pass word, must serve as a means of recognition. (See Darwin's " Variations of Animals and Plants," 1868, ii. p. 333.)

Huber (*loc. cit.*) remarks that ants from the same nest recognised and caressed each other with their feelers after four months' separation. If ants from two heaps of the same species meet in battle, the ants of the same side will seize each other in the general whirl, but recognise each other as soon as their antennæ have touched and mutually apologise.

Forel observed some distrust at first after a long separation, but it did not last long, and soon gave way to mutual understanding and mutual help. He once put a single ant from an old nest on the dome of a new, which he had made about a month before as an offshoot of the old. It was at once surrounded by more than fifty ants, which tapped it on all sides in so pressing a manner that it did not know which way to turn. But after the questioners, clearly satisfied with their investigation, had departed, others came up which repeated the same manœuvre, and so on without cessation. The strength of the poor victim of curiosity appeared about to give way when suddenly an ant, touched with compassion, offered it its mandibles, and when the former quickly seized them and rolled itself round, tried to carry it inside the nest. But the crowds of the curious blocked up the entrances and the worries and tappings of the poor creature did not cease. One shameless ant even tried to pull it away from its bearer, until the latter at length reached a little stopped-up hole and attained its object.

In an experiment made by Forel, Amazons recognized their slaves almost instantaneously after a four months' separation. But after a year, according to his experiments, such recognition no longer took place, and this is the more likely as the life of an ant seldom lasts more than a year, and after such a lapse of time the nest, therefore, is occupied by new inmates. Yet W. M. R. (*loc. cit.*) had ants in his possession which attained to an age of from two to three years. This was especially the case with queens. The same observer had the opportunity of establishing recognition between friend and enemy after a separation of thirteen months. The friend was recognised and welcomed, but the enemy was driven away. Lubbock also (*loc. cit.*) proved that recognition took place after a separation lasting more than a year, and that individuals lived for from three to four years.

CHAPTER XIV.

WARS AND BATTLES.

THE wars and battles of the ants are sometimes waged between different nests or different colonies of the same species, and sometimes between distinct species and genera. They are also murderous in a sense in which the slave hunts, aimed at a single object, are not, as a rule, and the slain, wounded and maimed are found in numbers no smaller than those in the bloodiest wars of men. The irritation of battle also is not less than the similar irritation in human struggles, and all the wild passions of human nature, such as lust of blood and of slaughter, cruelty, etc., appear on such occasions to be as roused and as real in the little ant as in the "crown of creation." The warriors sometimes become intoxicated with the heat of battle, until at last they are so mad that they forget all prudence and often sacrifice or permit themselves to be killed in quite a useless way. In such case infuriated warriors can generally only be pacified by a number of their comrades holding them by the legs and stroking them with the feelers until the fit of rage has passed over. An observation of this kind among the Amazons has already been mentioned.

Hauhart of Bâle ("Scientific Journal of the Scholars of the Bâle High School, III., 1825, No. 2) observed a regular battle between the black garden ants (*F. nigra*) and the small black ants (*F. fusca*), and described it as follows :—

The *fusca* had two nests and the *nigra* five small ones close together, twelve yards apart. At Whitsuntide at 10 o'clock in the morning a great movement was seen among the *fuscæ*. They marched out against the *nigræ* in long oblique lines of battle, with two separate troops on the advanced left wing and three on the right at some distance. The numerous black ants arranged themselves in a rather deeper order of battle, also with two separate wings. The armies grappled and fought at first with closed ranks, then divided into duels, while the wings stood opposite each other doing nothing. The struggle went on with embittered rage, feelers and legs were torn off, the opponents bit each other without mercy, and while the black ants helped each other and defended or carried away their wounded, the *fuscæ* left theirs to their fate. When the observer visited the field of battle two hours later the *fuscæ* had been vanquished and had disappeared; here and there only was a fugitive to be seen. The black ants had taken possession of the nests of their enemies, and were running eagerly about, to and fro, between these and their own nests. The excitement and fury during the struggle were so great that if any of the fighters were taken up they ran over the hand without biting and did not touch sugar laid before them.

Here for the most accurate and detailed accounts we must again thank Forel, who observed battles brought about both artificially and naturally between ants of the same and also of different species. Battles between ants of the same species often end with a lasting alliance, especially when the number of the workers on both sides is comparatively small. The wise little animals under such circumstances discover, much more quickly and better than men, that they can only destroy each other by fighting, while union would benefit both parties. Sometimes they drive each other out of their nests in a quite friendly way. Forel laid on a table a piece of bark with a nest of the gentle *Leptothorax acervorum*, and then put on it the contents of another nest of the same species. The last comers were by far the more numerous and soon possessed themselves of the nest, driving out the inmates. But the latter did not know whither to go, and turned back again. They were then seized by their opponents one after the other, carried away as far as possible from the nest and there put down. The oftener they came back the further were they carried away. One of the carriers arrived in this fashion at the edge of the table, and after it had by means of its feelers convinced itself that it had reached the end of the world, mercilessly let its burden drop into the fathomless abyss. It waited a moment to see if it had attained its object and then turned back to the nest. Forel picked up the ant which had fallen on the floor, and put it down right in front of the returning ant. The latter repeated the same manœuvre as at first, only stretching its neck further over the edge of the table. He several times reiterated his experiment and always with the same result. Later the two colonies were shut up together in a glass case and gradually learned to agree.

On the 7th of April, 1869, a battle at first artificially brought about between two colonies of *F. pratensis* was preceded by a number of skirmishes until towards nine o'clock a regular general battle began, which went on for a whole hour, on the same spot. They fought in large compact masses, which were always freshly directed towards a point midway between the two armies. Side attacks, as with the *sanguinea*, did not take place. Chains of from four to ten ants which had clung together and covered each other with poison were not unusual. Forel often saw, as did Huber, warriors of the same side furiously attacking each other, but they at once recognised and let each other go. Prisoners were dragged into the hostile nest and there killed. Meanwhile work within the nest suffered no interruption, and all the labors of peace went quietly on in the midst of war. Towards ten o'clock the van of one army had broken through that of the other, but the vanquished set themselves against a little wall which served as a natural line of defence, and which was made of dry twigs, leaves and plants. At the same time a great movement was perceptible on the dome of the nest of the beaten ants, and some of the ants were working busily away with their antennæ. Immediately afterwards new masses of warriors came out of all the openings of the nest and hurried to the aid of their vanquished comrades. But the enemy, which meanwhile had been carrying off hundreds of prisoners, also made fresh efforts and the battle reached its highest point. The field of battle was covered with warriors thickly pressed together, yet the original lines of battle could still be distinguished. Towards eleven o'clock the original victors were flung back, and pursued first to the original field of battle and then as far as their nest. The conquerors were here obliged to halt, for a new storm had gathered against them in the shape of a third colony of *pratensis* brought there by Forel. They fought with success also against this new foe, but were finally so tired that towards three o'clock in the afternoon they drew back and rested. Hundreds, perhaps thousands, of slain covered the field of battle; the most frequent sight was that of two dead enemies firmly locked together and with their mandibles biting into each other.

On the following day there were only individual collisions, but on the third day Forel reawoke the struggle, which ended two days later with the destruction of the last assailants, although the victors had also heavy losses. Forel often saw one of them maimed by a single bite through head or thorax from a larger opponent.

The battles of these same ants with the *Tetramorium cæspitum* (turf ants)—one of the strongest and most war-like species belonging to the genus *Myrmica*—are very frequent and violent; poison and sting therein play chief part, while with the giant ants (*Camponotus herculaneus*) terrible mutilations made by the mandibles are the rule. The latter ants also give peculiar alarm signals at the beginning of the fight. They not only touch each other very rapidly and vigorously with their feelers, but also strike the ground or the wood of the tree in which is their nest so sharply and quickly with their abdomens that an audible noise is made. All the species of *Camponotus* do the same. They also mutually assist each other, and this is usually fatal to the ant which has been seized by several of them.

The otherwise more blood-thirsty *sanguineæ* are less quarrelsome with one another. Forel tried to incite one colony of these against another by throwing down between them a number of *pratensis* pupæ. But each side busied itself with the costly harvest without injuring the other; only here and there did little bickerings break out, and a few were dragged to the hostile nest. The sense of the little creatures shamed the tempter! In another case the experiment resulted in an alliance. Only the Amazons from different nests would never form alliances, but fought to the uttermost.

Between different species war to the knife is the general rule. While battle between members of the same species is at first brought about with difficulty and becomes gradually more vigorous, it breaks out at once decisively and bitterly between different species. Every energy is set upon destroying the opposing party. Alliances scarcely ever take place, and are quite impracticable between distinct genera and species. On the other hand truces occur, when both sides are exhausted by protracted struggle. Forel observed two neighboring nests of the sanguine ants and the *pratenses* which every spring carried on severe battles, lasting for a whole day at a time, without either being able to win supremacy. After a few days, and when the surrounding ground was covered with the fallen, the affair ended with a truce, which lasted for the rest of the fine weather. A neutral territory was marked out between the two nests, which was not to be infringed by either party. But if one or several ants were placed on the hostile nest bitter war again broke out.

On the 17th of April, 1870, Forel placed a handful of black *pratenses* on a nest of the same species, but of a rather lighter variety or sub-species. A most infuriated battle followed from which only four or five of the black ants escaped. The rest were killed in an hour's space.

On the 12th of May, 1871, Forel saw a fight between the large brown *Myrmica scabrinodis lobicornoides* and a small *Myrmica scabrinodis* of a yellow brown shade. The battle began by the more numerous small ants taking prisoner some of the large ones which had approached their nest. Fugitives carried the alarming news to the nest of the latter and a general battle was the result. The large ants advancing in masses quickly broke through the order of battle of the small, freed the prisoners and put their enemies to flight. But the latter hid themselves in the unevennesses of the ground in the depths of which was their nest, and sought thence to do as much harm as possible to their foes. Forel saw one of the large ants seized by three of the smaller, and pulled into the depth of the nest through a small hole. The small ones taken prisoner by the large were killed or dragged half dead to the nest, the large ones making so much noise with their stings and mandibles that Forel could hear the crackling against the hard and ribbed thorax of their enemies. The army of the large ants remained for a moment on the field of victory while they tried to penetrate through the nest holes of the small. But the latter had so well stopped up their entrances that the attempt did not succeed. The whole battle had only lasted for a quarter of an hour.

One day Forel put two handfuls of *pratenses* of two different sub-species or varieties before a strong nest of *rufibarbis*. They were beaten back by the latter, but had scarcely escaped their jaws before they attacked each other most furiously.

The powerful *Camponotus* species have a peculiar method of fighting which enables them to face the Amazons. They raise themselves as high as possible on their hind legs, so as to prevent any seizure of the back, and hold their open jaws towards their enemy, bending their antennæ back. At the same time they curve the abdomen so as to be able to inject poison into the wound made. None the less many were killed by a few Amazons which Forel set against some of the *C. ligniperdus*, and he observed that some were regularly decapitated. At last a kind of truce was made until Forel brought some new *ligniperdi* which killed all the Amazons and allied themselves with those of their own race.

The *F. exsecta* or *pressilabris* also fights in a peculiar way, which is due to care of their small and very tender bodies. It avoids all single combats and always fights in closed ranks. Only when it thinks victory secure does it spring on its enemy's back. But its chief strength lies in the fact that many together always attack a foe. They nail down their opponent by seizing its legs and holding them firmly to the ground, while a comrade springs on the back of the defenceless creature and tries to bite through its neck. But if threatened the holders sometimes take flight, and so it happens that in battles between the *exsectæ* and the much stronger *pratenses* not a few of the latter are seen running about with a small enemy clutching their shoulders, and making violent efforts to tear the neck of its foe. If the bearer is then seized with cramp, the nervous cord has been injured. On the other hand if an *exsecta* is seized by the back by a *pratensis* it is at once lost.

The tactics of the turf ants resemble those of the *exsectæ*, three or four of them seizing an opponent and pulling off his legs. In similar fashion the attack of the *Lasius* species is chiefly directed against the legs of its enemies, three, four or five uniting in the effort. They understand barricade-fighting particularly well in their large well-built dwellings, and if it comes to the worst fly by subterranean passages. They are feared by most ants on account of their numerical superiority. Forel one day poured the contents of ten nests of *pratenses* in front of a tree trunk inhabited by *Lasius fuliginosus* (jet ant). The siege at once began; but the jet ants called in help from the nests connected with their colony, and thick black columns were at once seen coming out from the surrounding trees. The *pratenses* were obliged to fly, and left behind them a mass of dead as well as their pupæ, which last were carried off by the victors to their nests to be eaten.

Lasius niger, the black garden ant, a very wide-spread species, often fights with varying results with *cæspitum*, *fusca*, *flavus*, *sanguinea*, different *Myrmica* species, etc. Forel one day saw thousands of them besieging a nest of the *rufibarbis* without any definite result arising on either side. Huber observed a battle between two nests of the *Lasius flavus* (yellow ant) which stole each others' plant-lice.

McCook (*loc. cit.*, 29 Jan., 1878) saw a battle between two nests of the *Tetramorium cæspitum* lasting for nearly three weeks, which took place near a church standing between Broad Street and Penn Square in Philadelphia. He also established the fact that in such a fight friend and foe recognise each other with absolute certainty after communication by their feelers, however great may be the whirl or confusion of the struggle, and although on investigation of different warriors no difference is perceptible. According to his recorded observation the reason of this striking fact lies in the sense of smell among the *T. cæspitum*, while similar experiments with *Camp. Pennsyl.* yielded quite contrary results. Further, individuals of the before-mentioned Pennsylvanian mound-building ants were treated as enemies on their return home after McCook had dipped them only once in water, and here the momentary loss of their specific smell can clearly alone be in fault. Strange to say, the ants thus treated offered no resistance to the assaults of their friends, as though they were conscious of their involuntary mistake.

The most formidable of all the European ants is, according to Forel, the rather rare *Myrmica*, or *Myrmica rubida*, for it knows how to make special use of its sting and can render itself very unpleasant even to men. Its sting is almost more powerful than that of a wasp. Forel saw the *rubidæ* kill a whole sackful of *pratenses* in less than an hour without losing one of their number. An Amazon which attacked a *rubida* was killed in a few moments. A handful of *rubidæ* which he placed in the midst of a nest of the *rufa* held possession of the dome and kept in check the countless swarms of their enemies which did not venture to climb it. Forel never saw them draw back in a fight.

The *Myrmica scabrinodis* is not very warlike but is all the more thievish. It steals its booty right out the nests of its enemies, being better guarded against punishment than other species by its hard, leather-like coat of mail. Forel saw a *scabrinodis* on the dome of a *rufibarbis* nest first pretend to be dead, and then quickly steal off with the body of an insect which a *rufibarbis* had brought up and had let go for a moment. It is constantly at war with the turf ants and generally keeps the upper hand.

That the *Attæ*, or harvesting ants, fight with each other for the sake of plundering their several stores of provisions has already been mentioned above.

There are also some species which scarcely ever fight, either because of their peaceable nature or because their nests are too small for the array of hostile armies, such as *Myrmecina, Leptothorax, Stenamma*, etc. When they are attacked they try to save themselves by flight, and protect themselves by making nests which are as small as possible in very concealed and little frequented places.

The genus *Pheidole*, lastly, demands very special mention, for it is almost the only one among European ants which has that special class of neuters usually called soldiers. In Asia, Africa, and America this genus is much more widely spread than in Europe.

These soldiers, which are distinguished from their sisters by an enormously large head and very powerful mandibles, play just the same part in the ant state as soldiers do among men; that is, they do not work but only fight and defend their working sisters. Lespès denies this so far as the *Pheidole megacephala* (large-headed ants) are concerned—a very small, light yellow *Myrmica* species living in Southern France and Italy, and observed by him; for he saw the soldiers, which are much larger than the workers proper and which have a head from six to ten times as big, working just like their comrades. On the other hand, Heer ("The House-Ants of Madeira," Zurich, 1852,) observed that the soldiers of *Pheidole* or *Œcophthora pusilla* (found in Spain) when meat and dead insects were supplied as food acted the part of butchers, dividing it with their large mandibles into small pieces which the workers carried into the nest. Cocci and a species of beetle lived in their nests. Forel also noticed a special class of soldiers among the *Ph. pallidula*, although they marched with the workers and mingled with them. Both are incredibly brave and self-sacrificing. On the other hand the soldiers never take any part in domestic duties, but undertake the defence of the nest and its entrances against enemies from without. Forel kept a colony of this species captive for a long while, and never saw the soldiers working but only marching about.

A battle, brought on by Forel, between the *Pheidoles* and the turf-ants was at first begun very unwillingly on the part of the former, for the latter were far more numerous as well as stronger individually. A large number of *Pheidoles* were killed and as a rule remained clinging to the legs of their slayers after succumbing to a bite or a sting. But as the soldiers of the *Pheidoles* gradually came up, the aspect of affairs was altered. They carefully avoided being caught by the legs, and snapped at the backs of their opponents with their powerful mandibles, thus breaking the neck. If this manœuvre did not succeed and they were forced to fight hand to hand they were frequently overcome. If a turf-ant tried to penetrate into the nest, the soldier or guard at the entrance met it with such powerful blows from its mandibles that it lost its balance and was pulled into the nest by the worker-ants. The latter withdrew more and more from the contest, while the number of soldiers steadily increased, and the decimated enemies were wholly put to flight.

The fight between one of the soldiers of the *Pheidoles* and a *Crematogaster scutellaris* is very comic, the latter chiefly relying on its poison. It lets its abdomen wander over the head of the soldier, which struggles in vain to pull off one of its limbs, and which becomes more and more infuriated at the poison.

There are soldiers also in the genus *Colobopsis*, employed always in guarding the small holes of their carefully concealed nests, and therefore hardly ever quitting it. In one of these Forel found 450 workers, 65 fertile females, 45 males and 60 soldiers; in other nests there were comparatively more of the latter.

The soldier-class among European ants vanishes in comparison with those among their relations in non-European and tropical countries, wherein they far surpass their European brethren in size and strength. As a striking type of these may be selected a species belonging to the genus *Eciton*, the South American visiting, driver or foraging ant, whose habits much resemble those of the already described West African hunting or driver ants. According to Peters (*loc. cit.* p. 58) these animals come in endless files out of the desert and again disappear therein. On either side of the column, soldiers—distinguished by their large heads and jaws —run backwards and forwards, keeping the procession in order and on the right road. Their march is checked by nothing, not even by water. They fear nothing and attack the largest and the smallest animals with the greatest courage. The inhabitants are warned of their coming by the arrival of the ant-eating birds (*Grallaria* and *Formicivora*). And they are not unwelcome, for they do no harm to plants and destroy all injurious insects, reptiles and mammalia. So when the birds appear, the people quit their dwellings into which the ants enter from all sides. They penetrate everywhere, into all holes and corners in floors, walls, and roofs, and the whole house is soon cleared of all the insects injurious in the tropics, such as wasps, moths, mosquitos, millipedes, spiders and scorpions, and also of snakes, mice and rats; nothing escapes them, and this done, often after great losses, their army proceeds on its way.

An eye-witness, Herr H. Kreplin of Heidemühl (Station Ducherow) who lived for nearly twenty years in South America as an engineer and had often the opportunity of seeing the driver ants in the virgin forests there, writes as follows to the author on May 10, 1876:—

"The first sight of this nation on the march is striking to those accustomed to observe as well as to ordinary workers. The train moves forwards in a column of two or three inches in width, with a regularity and order which is astonishing when we consider the length of the procession and the extraordinary difficulties of the forest ground. If the travellers are more closely scanned they will be found to be of different sizes and colors. The ants marching in procession are about 7mm. long and are dark brown. They carry the larvæ (pupæ?) of the household fast gripped by

the stomach, and in spite of this burden move easily and rapidly. On both sides of the train, at about 10mm. distance from each other, stronger ants are to be seen, distinguishable from the others by their foxy color and very thick heads with gigantic mandibles. These "thickheads" play the same *rôle* in the ant-state for which they are cast in cultured communities. They look after the order of the march, and allow none to turn either to the right or left. The least confusion in the regularity of the march makes them turn round and put things straight again. While the procession of the brown workers streams on unceasingly with a swarming motion, the "officers," as the natives call these thickheads, run constantly backwards and forwards, ready to take the command on meeting any difficulty. The crossing of streams by these creatures is the most interesting point. If the watercourse be narrow, the thickheads soon find trees, the branches of which meet on the bank on either side, and after a short halt the columns set themselves in motion over these bridges, rearranging themselves in the narrow train with marvellous quickness on reaching the further side. But if no natural bridge be available for the passage, they travel along the bank of the river until they arrive at a flat sandy shore. Each ant now seizes a bit of dry wood, pulls it into the water and mounts thereupon. The hinder rows push the front ones even further out, holding on to the wood with their feet and to their comrades with their jaws. In a short time the water is covered with ants, and when the raft has grown too large to be held together by the small creatures' strength, a part breaks itself off and begins the journey across, while the ants left on the bank busily pull their bits of wood into the water and work at enlarging the ferry-boat until it again breaks. This is repeated as long as an ant remains on shore. I had often heard described this method of crossing rivers, but in the year 1859 I had the opportunity of seeing it for myself at the junction of the large River Gaspar (?) which falls into the———(?)
.... I never found the visiting ants seize provisions, although they have often driven me out of my home. But everything which flies and crawls is destined for their prey, and a house visited by them is thoroughly cleansed from all vermin. If they change their marching order for hunting order in the forest the wood becomes lively. All things,

even snakes, fly before the advancing foe. Just as little have I had the opportunity of establishing the carrying off of stranger larvæ to bring up as workers, although I have been struck with a difference in size and color."

We have here again to thank the English traveller Bates —from whom we have already borrowed the account of the Brazilian *Sa-ubas*—for the most complete accounts of these remarkable creatures. The *Sa-ubas* are often confused with the *Ecitons*, although their habits are very different and although the two species belong to quite distinct groups of ants. The Indians, who are very careful not to be caught by these foraging-ants when journeying through the woods, call them the *Tauocas*. Bates learned to distinguish ten distinct species of which eight were before unknown, each species having a special way of marching. In Ega [Brazil, near the Amazon.—Tr.] the woods swarm with their hosts. An interesting meeting of two armies of *Sa-ubas* and an *Eciton* species (*E. Canad.*) in Guiana near the Sinamari river is described by Var (in Brehm, "Animal Life," IX., p. 269).

The contrast between soldiers and workers, or workers major and workers minor, as Bates names the two classes, varies much in different species, and is greatest in the species *E. hamata, erratica,* and *vastator*, while, among other species (as *E. rapax, E. legionis,* etc.) the soldiers work just the same as the ordinary neuters. All the *Eciton* species are hunting animals, and they all hunt together in large armies, although each species has its own particular method. *E. rapax*, the soldiers of which are half-an-inch long, hunts in small troops and marches in small trains through the woods to rob the nest of another ant of the genus *Myrmica*. Bates often saw the mutilated bodies of the latter dragged away by the robbers. *E. legionis* also robs the ants of other species, and drags its slain enemies home to devour, after it has divided their bodies into two or more pieces, they being too heavy for a single robber to carry. In dragging their enemies out of their mined passages Bates saw how some dug a shaft, while others stood above to take the dug-out earth from their companions and carry it away far enough to prevent it from rolling again into the shaft. Here we again find that division of labor which appears everywhere to be a principle of work among the industrious ant nation, and has doubtless much contributed to the perfection of their customs and arrangements. On the return of a robber ant army to its dwelling, Bates also observed that the unladen ants helped their laden comrades to climb a steep wall.

The most common species are *E. hamata* and *E. drepanophora*, which traverse the forests along the banks of the Amazon in thick columns of countless thousands. The first sign which warns the foot passenger of their approach is the restless fluttering round of a number of ant-eating birds. If he overlooks this sign and goes on for a few steps, he is certain to find himself suddenly seized by thousands of tiny furious creatures, biting and stinging as hard as they can. Rapid flight is the only means of salvation, and the ants have to be pulled off the skin one by one, often leaving head and mandibles in the wound.

All living things, therefore, that are within their reach fly unresistingly out of their way. Unwinged insects, such as spiders, caterpillars, crickets, larvæ, other ants, etc., have the gravest reasons for escaping; only the birds and their broods are safer, for the *Eciton* does not love climbing tall trees. The main column, four to six deep, marches constantly forward in a given direction, clearing its road of all living or dead animals as it goes, and sending out from time to time small side columns, which return to the main body after completing their plundering. If a specially good supply of prey is found, as for instance a heap of rotten wood with many insect larvæ, a halt is made, and all that is edible is devoured. When they seize wasps' nests, which are sometimes placed on low bushes, they gnaw off the coverings of the larvæ, pupæ, and new-born wasps, and tear them all in pieces, untroubled by the furious owners flying round them. When the booty is carried off, the various pieces are apportioned according to strength; the small ants take the smallest and the large the heaviest loads. Sometimes two ants combine to carry a large burden. The soldiers, or workers major, with their heavy distorted jaws, alone take no share in this labor. The army never follows a trodden road but marches through the most impenetrable underwood. Bates never saw them turn back, but always on the march, and he failed to find any of their nests.

One day, at Villa Nuova, Bates saw at a favorable spot one of these columns sixty or seventy yards in length,

without either vanguard or rearguard. Order was maintained instead by single ants, running constantly backwards and forwards on both sides of the column, keeping up a kind of mutual understanding. These " officers " were seen to communicate with those marching in rank by touching them with their feelers. When Bates broke the procession or took an ant away the news of the fact reached the end of the army very rapidly and a retreat began. All the small-headed workers carried in their mouths pieces of white crickets whose nests they had plundered. The large-headed ants, whose shining white heads made them easily recognisable, never carried anything, but, as already said, ran outside the procession, like subalterns in a marching regiment and at regular distances from each other. Yet it appeared to Bates as though they were less warlike than their working comrades, and owing to their thick heads and curved mandibles they were also less active. Perhaps they only act as directors or overseers. Perhaps also as riding horses, if an observation of Bastian (" Travels," 294) quoted by Perty be accurate; Bastian says that he saw in Siam an army of black ants accompanied by soldiers, and that some of the former now and then left the ranks, sprang on the backs of the much larger soldiers, and trotted up and down the procession on them like officers, after which they again fell into the ranks! ? ?

The *Ecitons* are not always working and marching; they also take rest and refreshment. They halt on sunny spots in the forest, clean themselves or each other by wiping their antennæ with their fore-legs, or drawing their antennæ and legs through mouth and mandibles, and then walking slowly about or playing with each other like young lambs or puppies.

Eciton predator, a small black-red species, very common round Ega, does not hunt in columns but in thick masses, composed of myriads of ants, looking like a stream of dark-red liquid. They search every spot on their road most closely for animal food, and tear their victims in pieces in order to carry them away. Their armies often occupy a space of from four to six square yards, and single groups break off from the sides, like the skirmishers of sharpshooters of a human army, returning to the main body when they have obtained their object.

There are some blind or half blind species which avoid light, and when their road leads them over open places they cover it in with remarkable swiftness by building tunnels or galleries of earth. Some, such as *E. vastator* or *E. erratica*, march only by these covered ways. Bates was able to follow these roads for many hundred yards, and they were built in the same way as the covered ways of the Termites, only with the difference that the latter use a gummy saliva for glueing the earth together, while the ants simply heap it up so skilfully that it does not fall, although lacking any kind of mortar. The large-headed ants here act only as soldiers, defending the community against all disturbance from without, as amongst the Termites. When Bates made a breach in the covered ways, the small-headed ants tried to mend it as quickly as possible, while the large-headed ones rushed out menacingly, furiously snapping with their mandibles. The species observed by McCook in North America, such as the Texan cutting ants, the *Camponotus Pennsylvannicus*, and the mound building ants in the Alleghanies belong, according to his observations, to the soldier species, to those, namely, which keep a standing army.

The military condition, however, is most perfectly developed among the so-called white ants, or Termites, living in Africa, Southern Asia, South America, and Australia; these maintain just as numerous and well-disciplined an army as do our large European military powers. Yet their finances do not suffer as much thereby, as do those of human States, nor are their swashbucklers guilty of excesses against the citizens who feed them and whom they ought to defend. Do not be angry, dear reader! They are only unreasoning creatures, following mere "instinct," and cannot rise to the height of human perfection.

THE TERMITES; OR WHITE ANTS.

CHAPTER XV.

TERMITES AT HOME.

THE Termites are wrongly named ants, for they belong to an entirely different order of the *Insecta*, the *Orthoptera*, and are related most nearly to our *Blattæ* or cockroaches. They are three or four times as large as our common black ants, but are unfortunately far less accurately known. Their polity seems to be almost more developed than that of the ants, and their architectural talent is also superior. They raise, in Africa at least, fine buildings of from ten to twenty feet high, out of earth, clay, pieces of plants, stones, etc., fastening together these materials by a kind of gummy saliva. So firm does this make their nests, built in the shape of a cone or of a large haycock, that several men can stand on their surface, and that antelopes and even buffaloes are wont to use them for sentries to look over the wide plains. They do not even break through under the tread of an elephant or the weight of a heavily laden wagon. In Senegal their size and number are often so large that at a distance they resemble human dwellings, the similarly conical huts of the negro-villages, and travellers are often thereby led in a wrong direction. Jobson, in his "History of Gambia," says that many of these heaps are twenty feet high, and that he and his companions often hid behind them when out hunting. At first the buildings are only small, and resemble pyramids scarcely a foot high. Gradually, as the population increases, new and similar hills rise up all round. The partition walls are then broken through, the new dwellings are united to the old, a dome is added, and a symmetrical roof is built over all. This is continually repeated, until the mound of twelve or twenty feet high is made. The outer covering consists of a firm domed vaulted layer of clay, which is exceeding strong, so as to withstand injuries from weather, attacks of enemies, and other accidents.

The outer forms of the Termites' hills vary in different species. While most are conical, others resemble blunted pillars or giant fungi, the latter having domed roofs, overhanging five centimetres all round, and resting on a tall cylindrical support four or five feet high. In places subject to great and regular inundations, the Termites' nests are found barrel-shaped and built on the gnarled branches of strong trees, with tubular passages running down the trunks to the ground. Some species live in decayed trees, others subterraneously.

The ground round each Termites' nest is perforated for a considerable distance with tunnels as much as twelve inches wide, so as to keep up circulation and communication among the inhabitants. There is also a well-organised system of surface and subterranean canals and drains, so as to protect the building from the effects of the waterspout-like torrents of rain common in the tropics.

According to Bastian ("The Nations of Eastern Asia," II., p. 293) the Termites' towns of Burmah and the neighbouring countries are as high as a man, and often resemble a regular castle, with pinnacles and towers, while others again are simple massive tumuli or mounds. They seemed to him to be generally built round a decayed tree trunk.

If we compare the size and the extent of these buildings with the size of the builders, every work of man sinks into insignificance. A pyramid, on the same scale, would have to reach the enormous height of nearly 3,000 feet, and a subterranean passage, similar to a Termites' tunnel, would have a diameter of 300 feet. Yet we marvel at the Roman cloacas or the American aqueducts because a man can stand upright or sit on horseback therein.

The astonishment felt at the capabilities of these creatures —which Blanchard (" Rapport sur les Travaux Scientifiques des Départements en 1868") calls a scourge of the inhabitants of the countries in which they live but one of the wonders of creation for students of nature—becomes even greater when we investigate the interior of the hills which serve as their dwellings, about which unfortunately we have only as yet very imperfect information. These internal arrangements are so various and so complicated that pages of description might be written thereupon. There are myriads of rooms, cells, nurseries, provision chambers, guard-rooms, passages, corridors, vaults, bridges, subterranean streets and canals, tunnels, arched ways, steps, smooth inclines, domes, etc., etc., all arranged on a definite, coherent, and well-considered plan. In the middle of the building, sheltered as far as possible from outside dangers, lies the stately royal dwelling, resembling an arched oven, in which the royal pair reside, or rather are imprisoned, for the entrances and outlets are so small that although the workers on service can pass easily in and out, the queen cannot, for during the egg-laying her body swells out to an enormous size, two or three thousand times the size and weight of an ordinary worker. The queen, therefore, never leaves her dwelling and dies therein. Round the palace—which is at first small, but is later enlarged in proportion as the queen increases in size until it is at least a yard long and half a yard high—lie the nurseries, or cells for the eggs and larvæ; next these the servants' rooms, or cells for the workers which wait on the queen; then special chambers for the soldiers on guard, and, between these, numerous store-rooms, filled with gums, resins, dried plant-juices, meal, seeds, fruits, worked-up wood, etc. According to Bettziech-Beta, there is always in the midst of the nest a large common room, which is used either for popular assemblies or as the meeting and starting point of the countless passages and chambers of the nest. Others are of the opinion that this space serves for purposes of ventilation.

While the magazines, or storerooms, are built of clay, the nursery-cells are entirely made of woody material, fastened together with gum. In these are the eggs, and the newly hatched animals or larvæ, which are fed by the workers until they are grown large and can help themselves. These nursery-cells lie all round the royal cell, and as near to it as possible, so that the eggs can be easily and quickly carried into them, and their number is increased in proportion as the queen enlarges and lays more eggs. The servants' rooms are also increased as the care of the queen and the distribution of the eggs in the nurseries demands more and more service. Since, as already mentioned, the royal cell itself has to be enlarged, and as for this purpose the surrounding rooms have to be continually broken down and rebuilt further off, a constant feverish activity prevails in the interior of the nest, and all tasks are performed with wonderful acuteness, regularity, and prudence.

Above and below the royal cell are the rooms of the workers and soldiers which are specially charged with the care and defence of the royal pair. They communicate with each other, as well as with the nursery-cells and store-rooms, by means of galleries and passages which, as already said, open into the common room in the middle under the dome. This room is surrounded by high, boldly projected arched ways, which lose themselves further out in the walls of the countless rooms and galleries. Many roofs outside and in protect this room and the surrounding chambers from rain, which, as already said, is drained away by countless subterranean canals, made of clay and of a diameter of ten or twelve centimetres. There are also, under the layer of clay covering the whole building, broad spirally winding passages running from below to the highest points, which communicate with the passages of the interior, and apparently, as they mainly consist of smooth inclines, serve for carrying provisions to the higher parts of the nest.

It is exceptionally difficult to investigate accurately the interior of a Termites' nest, owing to the interdependence of the several parts—the destruction of one room, arch, or passage causing the breaking down of many; added to this, the energetic resistance of the Termites' soldiers, armed with very sharp and strong mandibles, puts great obstacles in the way of the observer. "They fight," says the English traveller Smeathman, the distinguished Termites' observer, to whom we owe the best and fullest information upon these creatures, "They fight to the last man, and they defend so energetically every inch of their property that they often drive away the unshod negroes, while the blood of the European runs through his stockings. We were never able to study the interior of a nest in peace, for while the soldiers attacked us, the workers stopped up as quickly as possible the rooms and passages laid open. They do this especially in the neighborhood of the royal dwelling, for which they show the greatest care, and that so cleverly that from outside it only looks like a formless heap of clay and cannot be distinguished from its surroundings. Nevertheless it is not hard to find, partly from its situation in the midst of the building, and partly because it is surrounded by great crowds of workers and soldiers, willing to risk their lives in its defence. The interior also, in addition to the royal pair, is found filled with hundreds of the workers serving the latter. These faithful servants do not desert their sovereigns even in utmost need and peril. For when I," says Smeathman, " took out such a royal dwelling and kept it in a large glass vessel, all the servants busied themselves with the greatest care about their sovereigns, and I saw some of them engaged at the head of the queen as though they were giving her something. They then took away from her abdomen the eggs laid by her, and carried them carefully into some unbroken parts of the building, or between scraps of clay as well as they could."

Life in such a nest, and especially the remarkable division of labor between workers and soldiers, can best be observed by making a sudden attack on the mound. Smeathman and other observers, such as Forskal, König, Sparman, etc., say that if a hole is made from outside in a Termites' hill with a strong hoe or axe the first thing that catches our attention is the behavior of the soldiers. Immediately after the blow a single soldier (perhaps a general or one of the higher staff officers?) appears in the breach, and seems to seek for the cause of the injury and the nature of the foe. He then withdraws into the interior, and gives a signal of alarm, whereat in the shortest possible time, as quickly as the size of the hole permits, masses of soldiers pour out. It is difficult to describe the fury with which these warlike insects fight. In their eagerness to drive back the enemy they often fall down the sides of the mound, but they soon recover themselves and bite at everything that comes in their way. This snapping, together with the striking of their mandibles against the building, make a crackling or tremulous noise, rather sharper and quicker than the ticking of a watch, which can be heard at a distance of several feet. During the attack they are in the most active movement and excitement. If they succeed in reaching any part of the human body they instantly inflict a rather painful wound, and a spot of blood more than an inch in size appears on the stocking. Their curved mandibles meet at the first bite, and do not loose their hold even when the creature's body is torn off bit by bit. On the other hand, if the assailant withdraws beyond their reach and inflicts no further injury, they retire within their dwelling in the course of half-an-hour, as though they had come to the conclusion that the enemy who had done the mischief had fled. Scarcely have the soldiers disappeared when crowds of workers appear in the breach, each with a quantity of ready-made mortar in its mouth. As soon as they arrive they stick this mortar round the open place, and direct the whole operation with such swiftness and facility that in spite of their great number they never hinder each other, nor are obliged to stop. During this spectacle of apparent restlessness and confusion the observer is agreeably surprised to see arising a regular wall, filling up the gap. During the time that the workers are thus busied the soldiers remain within the nest, with the exception of a few, which walk about apparently idly, never touching the mortar, among the hundreds and thousands of workers. Nevertheless one of them stands on guard close to the wall which is being built. It turns gently each way in turn, lifting its head at intervals of one or two minutes to strike the building with its heavy mandibles, making the before-mentioned crackling noise. This signal is immediately answered by a loud rustling from the interior of the nest and from all the subterranean passages and holes. There is no doubt that this noise arises from the workers, for as often as the sign is given they work with increased energy and speed. A renewal of the attack instantaneously changes the scene. "At the first stroke," says Smeathman, " the workers run into the many tunnels and passages which run through the building, and this happens so quickly that they seem regularly to vanish. In a few seconds they are all gone, and in their stead appear the soldiers once more, as numerous and as pugnacious as before. If they find no enemy, they turn back slowly into the interior of the hill, and immediately the mortar-laden workers again appear, and among them a few soldiers, which behave just as on the first occasion. So one can have the pleasure of seeing them work and fight in turn, as often as one chooses; and it will be found each time that one set never fight, and the other never work, however great the need may be."

Quatrefages ("Souvenirs d'un Naturaliste," II., p. 405) also never saw the soldiers working, but only acting as leaders or overseers. Fritz Muller also—who lately published some interesting observations on South American Termites, especially on the species named by him the *Termes Lespesi*, and on the round nests built in trees by some species—describes things as happening just in the way related by Smeathman. If a piece of the nest is broken off, the workers withdraw from the uncovered passages, while the soldiers appear in large numbers in their place, and run hither and thither, continually touching each other with their feelers. After some time the workers return and eagerly toil at stopping up the openings, partly with earth, partly with their own excrement. Meanwhile the soldiers have withdrawn into the interior, with the exception of a few which appear to overlook and encourage the workers. If a grass-stalk is held in the uncovered passages of a Termites' hill, as the natives do for the sake of catching them, the soldiers will bite it and will let themselves be pulled out thereby.

If the Termites excel in building nests, they excel yet more in making roads and bridges, for their proceedings here really touch the fabulous. All their roads are subterranean or covered, for they either shun the light or else seek thus to withdraw themselves from the eyes of their numerous enemies. Perhaps also they wish to escape from the burning rays of the sun. "Wherever they go," says Dr. A. Hagen (" On the Habits and Distribution of Termites," 1852), " and however far it may be, they first build a viaduct, a tunnel as thick as a quill pen, made out of clay, smooth inside and more or less rough outside. It is wonderful how rapidly their work progresses. Marching in closed ranks, each worker brings to the proper place a little pellet of earth mixed with saliva, secreted by its large glands. Its strong head seems to serve as trowel and hammer. All observers agree that the tunnels seem to grow before their eyes, almost imperceptibly, and Forskal relates that the Termites watched by him in Egypt built two inches of a tunnel in an hour, and three yards during the night. The little crowd toil without ceasing at the enlargement of their work." In order that the labor may be continual, troops of workers apparently relieve each other. When it is practicable they tunnel beneath the earth, but can work quite as well in the open when circumstances demand it. If for instance they come across a rock which they cannot perforate, when they are making an underground road, they build a tubular passage over it. They can even carry their viaducts through the air, and that in such bold arches that it is difficult to understand how they were projected. In order to reach a sack of meal which was well protected below, they broke through the roof of the room in which it was, and built a straight tube from the breach they had made down to the sack. As soon as they tried to carry off their booty to a safe place, they became convinced that it was impossible to pull it up the straight road. In order to meet this difficulty, they adopted the principle of the smooth incline, the use of which we have already seen in the interior of their nests, and built close to the first tube a second, which wound spirally within, like the famous clock-tower of Venice. It was now an easy task to carry their booty up this road and so away.

"Like clever engineers," says Blanchard (*loc. cit.*) " they build tubular bridges to pass from one spot to another, or tubes from one story of a building to the next. In the cellars of the prefecture of La Rochelle (Southern France) a number of hollow pillars were to be seen, like stout straws, reaching from floor to ceiling." In building they always cling to the principle that the shortest way is the best, and it is marvellous with what certainty they take the straightest road, even underground, to their places of supply. It has been thought that during the night they send out scouts, which search for the road above ground, and by given signals point out the direction to their comrades working beneath. After what we have already seen among ants, and shall see among bees of the habit of sending out scouts, this does not seem wholly improbable.

It has already been said that their exists round their nests a widespread system of subterranean canals, serving as means of communication between neighboring colonies. These canals are the wider, the nearer they are to the nest itself, where they have often a diameter of half a foot or more, and become narrower as they get further off—thus exactly agreeing with the principles of regular road and canal making.

Let us now throw a rapid glance at the remarkable insect itself; its housekeeping requires no more especial description, as it closely resembles that of the ants already described, and only seems more complicated owing to the greater number of differing individuals. Lespès has shown that in the nests of the small Termites' species (*Termes lucifugus*) found in France, in addition to the larvæ of males, fertile females, and neuters, and in addition to the workers and soldiers of the latter, there are nymphæ of two other kinds; the smaller with short and the larger with longer wing-rudiments. There are also two kinds of males and of fertile females. The smaller ones appear towards the end of the month of May; the others, much larger, only in August. Lespès calls the first the minor, the latter the major, kings and queens. The most remarkable are the soldiers, among which Lespès has recognised the barren of both sexes, and of these there is about one to a hundred workers—a far better ratio than among men, whose standing armies are often a fiftieth or a thirtieth of the total population. The soldiers have enormously large, hard, and strong heads, almost as big as the rest of their bodies, and these are armed with gigantic and very strong and sharp mandibles, while the heads and mandibles of the non-combatant workers are much smaller and weaker; the whole body of the latter is also of less size and strength. Their mandibles are suited only for gnawing and grasping, whereas those of the soldiers serve as terrible weapons. The part played by the soldiers in defending the nest has already been mentioned, and they also appear to act as overseers and directors. It was told above how they watched over the repair of the breach. Smeathman saw further one day, as he was passing through a forest in West Africa, a large army of the so-called marching Termites, whose larvæ and soldiers are not blind, as in the other species, come out of a hole in the ground, and disappear again into a similar hole at some distance. Their number was very large and they marched with the greatest swiftness in thick ranks of fifteen, mostly workers. Here and there Smeathman saw a soldier marching in the same way, carrying his heavy head with apparent difficulty. One or two feet off the column appeared other soldiers, either standing still or walking up and down as if watching lest any enemy should threaten a surprise. Others had climbed up neighboring plants, looked around, and made the already-mentioned crackling noise, whereupon the whole army replied in like fashion, quickening their steps. Smeathman watched them for more than an hour without perceiving the least decrease of strength in the procession.

The most important personage of the Termites' State is obviously the queen, since on her existence depends the existence of the whole nation. If the royal cell is removed from a Termites' nest, the colony scatters or perishes. On the other hand, the whole of the building can be destroyed without any such result, provided only the royal cell is left standing; the remainder will be rebuilt. When the queen dies the community would perish, were it not that the wise insects prepare for such an emergency by keeping some reserve-queens. "In each nest, in a little dwelling resembling the royal cell, are two or three expectant queens, which only receive their investiture on the death of the queen-mother, and then begin to provide for the welfare and the increase of the nation" (Hagen, *loc. cit.*).

The queen lays in her cell an enormous number of eggs, often as many as 80,000 in twenty-four hours, and these are at once taken away by the serving workers and carried into the surrounding cells. "An endless stream of workers moves round the floor of the royal cell, carrying the eggs into the nurseries around. To shorten the journey little holes are broken through all round at regular distances, and are used for short cuts by the laden workers. The eggs themselves, differing according to the size of the species, resembling either powdered sugar or the so-called ant-eggs, are stored in the already-mentioned cells—which have been termed lying-in rooms—one over another. All kinds of food have been carried into special magazines for the first nourishment of the young when hatched, and there are soon representatives of all stages of development mixed confusedly together; it is a whirlpool of shapes, forms, and colors, which, however, duly leads to a single species" (Hagen, *loc. cit.*).

The wedding flight of the Termites' males and females resembles to a hair that of the ants. Hagen *(loc. cit.)*, at once accurately and poetically, describes it in the following fashion:

"Let us place ourselves in a forest spot in the interior of Brazil. Not far from a whispering brook a clearing begins, the thicket opens and hems in a vale covered with underwood. Here and there several mounds of earth rise foot-high, turfed above and not unlike gigantic mole hills. A pleasant resting-place for the weary traveller, cared for by heaven, for the end of his journey is yet distant and the sun already near the horizon. A thick dark cloud floats slowly overhead and renders the close evening yet more sultry. The rainy season, the weary winter of the tropics, is at hand. The staff is already seized to hurry with freshened energy to the hospitable hearth, when the parting glance thrown by the traveller on the place of repose is suddenly arrested, for an uneasy movement begins in the seemingly dead mound of earth. As by enchantment a diagonal slit opens in the midst of the hill. A little insect squeezes itself out, with inch-long winglets closely pressed together: it is followed by two, three, four and more in a row, as many as can pass through the rapidly widening crevice. The band descends the hill like a silver ribbon, the delicate membrane of the thousands of tiny wings glittering like mother-of-pearl. The procession makes its way exactly against the wind, for thus only can the tender winglets resist the pressure of the air. Swiftly, without ceasing, it continues, new and ever new arrivals strengthen the troop as swiftly as if they were being driven from the nest. Meanwhile more rifts have been made. Similar swarms break forth from these. Like a volcano the tiny mountain seems to discharge its living lava. At the rifts themselves is seen a curious spectacle. Tiny unwinged creatures with uncouth heads and sabre-like curved jaws appear at the openings. Threateningly they swing their large heads, and defend the entrance to their subterranean chambers or hasten the march of their expelled brethren. The wonderful procession lasts for a full hour; it almost seems as though it were to have no end. At last the ranks become thinner and narrower, here and there are seen delaying laggards, the crevices, walled in by unseen hands, begin to close and the hill soon takes on its wonted appearance. Meanwhile the troop has taken wing; it gradually rises with uncertain wavering flight higher and higher, and hovers closely pressed together round the tops of the trees. A continual rising and falling of individuals enlivens the scene and changes the whole whirl into that mystic dance which the ephemeral gnats are wont to lead here also on the warmer summer evenings. Gradually the number of the falling insects increases. If we look closely

o

we always find them in pairs; a larger one is closely chased by a smaller and seized in its jaws. Then both run round quickly, and try to pull off their loosely attached wings with their feet. The picture is also enlivened in another fashion. Countless insectivorous quadrupeds, birds, lizards, snakes, and frogs have gathered. The unarmed Termites, now incapable of flight, are swallowed in masses, and even man finds the food a dainty. If we follow yet further the life-path of these scarcely-born creatures, we shall find few of these myriads alive on the following morning. Those which did not fall a prey to the rage of hunger, wander about shelterless, or are caught by the Termites' workers, now busily creeping about, and are selected for the future heads of families," etc. All the males and fertile females which are not chosen and protected in this way are fated to perish. "The manner," says Smeathman, "in which the workers protect the fortunate elected pair from their numerous enemies, not only on the day of the general massacre but for a long time afterwards, justifies my expression 'election.' The little busy creatures at once shut up their chosen in a room of clay, which at first has only a single small entrance, which lets pass themselves and the soldiers, but not the royal pair. Later on several entrances are made, but always so that the care of defence and nourishment is left to the people alone."

CHAPTER XVI.

TERMITES ABROAD.

THE Termites show even more intelligence in their activity abroad than in their home or family life, and this activity makes them one of the heaviest and most dreaded of scourges in the countries they inhabit. They are born destroyers, and spare nothing that is not either of stone or iron. Especially are all wooden things subject to their attacks, and their inroads are the more dangerous because they are not visible to the eye, and are, as a rule, first discovered when it is too late to hinder them. Either from the desire to remain undiscovered, or from their liking for darkness, they have the remarkable habit of destroying and gnawing everything from within outwards, and of leaving the outside shell standing, so that from the outside appearance the dangerous state of the inside is not perceptible. If, for instance, they have destroyed a table or other piece of household furniture, in which they always manage from the ground upwards to hit exactly the places on which the feet of the article rest, the table looks perfectly uninjured outside, and people are quite astonished when it breaks down under the slightest pressure. The whole inside is eaten away, and only the thinnest shell is left standing. If fruits are lying on the table they also are eaten out from the exact spot on which they rest on the surface of the table.

In similar fashion things consisting wholly of wood, such as wooden ships, trees, etc., are destroyed by them so that they finally break in without anyone having noticed the mischief. Yet it is said that they go so prudently to work in their destruction that the mainbeams, the sudden breakage of which would threaten the whole building and themselves therewith, are either spared, or else so fastened together again with a cement made out of clay and earth that their strength is greater than ever! (?) Hagen also states that they never cut right through the corks which stop up stored bottles of wine, but leave a very thin layer which is sufficient to prevent the outflow of the wine and the consequent destruction of the workers. The same author relates that in order to reach a box of waxlights they made a covered road from the ground up to the second story of a house.

The Termites were first introduced into Europe by a ship from over the seas, and have made themselves remarkable as the most mischievous enemies of wood in Italy, Spain, France, and the greenhouses in Schönbrunn, near Vienna. In France they have settled along the banks of the Lower Charente, in the towns of Rochefort and la Rochelle, and also in Bordeaux and the vicinity. They were in Rochefort apparently for a long time before they were discovered, until the fall of an uninhabited house in the Rue Royale and the simultaneous enormous dispersion of the Termites in the neighboring houses, in the year 1797, drew on them the eyes of the authorities, unfortunately too late. Closer investigation showed that the whole costly stores of stacked oak for the building of ships of war for the navy had been destroyed; all the public buildings were infected and the archives of the navy could only be thenceforth protected by keeping them in metal safes. In a school a whole dinner suddenly sank two stories deep into the cellar, and other buildings threatened to fall. A smith, living in the neighborhood of the docks, suddenly saw his anvil yield under the strokes of his hammer. The wooden block which carried it splintered in pieces, and revealed itself as the dwelling of the Termites. In the year 1820 the ship of war, *le Génois*, built under Napoleon, had to be broken up, being rendered quite useless by the Termites. The same fate befell an English ship of the line, the "Albion," into which the Termites had penetrated.

The Termites apparently came to Schönbrunn in plants from Brazil. They so thoroughly destroyed both the wooden tubs and the beams that in the year 1839 one of the largest greenhouses had to be pulled down. They multiplied rapidly within the greenhouses, at a temperature of $24°$ R. [$30°$ C., or $86°$ F.—Tr], but are now nearly exterminated. The European Termites belong almost exclusively to the already mentioned *Termes lucifugus* (light avoiding Termites), on which the detailed observations of Lespès were made, and in whose habits he found many differences from the non-European species.

According to Blanchard the third part of the plains of the Island of Ceylon is undermined by the Termites. In Upper Egypt they not seldom compel the inhabitants to leave their ruined dwellings and to build a home in a new spot. In the East Indies, Bengal, the southern part of China, Soudan, etc., they are a fearful scourge. In West Africa they level to the ground in a few years the deserted huts of the natives. In the whole of South America, as Humboldt relates, books more than fifty years old are rare, for the Termites have the laudable habit of making passages into the libráries, and obliquely through the rows of books. In the sea-coast towns of Brazil and of the East Indies whole magazines often fall a prey to their destructive energy. Even metal is not secure from the attack of the sour Termites'-acid, and the bores of iron cannon in Ternate [Molucca Isles] are found covered with Termites' roads and are quickly attacked by rust.

On the Termites of South America the English traveller, Bates—who has given us so many interesting details on the ants of that continent—has again made a report which contains indeed nothing new, but which deserves to be given here as the record of the observations made lately by a trustworthy eye witness:

"The surface of the Campos" (round Santarem, a town lying on the lower course of the Amazon), says Bates, "is disfigured in all directions by earthy mounds and conical hillocks, the work of many different species of white ants. Some of these structures are five feet high, and formed of particles of earth worked into a material as hard as stone; others are smaller, and constructed in a looser manner. The ground is everywhere streaked with the narrow covered galleries which are built up by the insects of grains of earth different in color from the surrounding soil, to protect themselves whilst conveying materials with which to build their cities—for such the tumuli may be considered—or carrying their young from one hillock to another. The same covered ways are spread over all the dead timber, and about the decaying roots of herbage, which serve as food to the white ants. An examination of these tubular passages, or arcades, in any part of the district, or a peep into one of the tumuli, reveals always a throng of eager, busy creatures. .... White ants are small, pale-colored, soft-bodied insects, having scarcely anything in common with the ants, except their consisting, in each species and family, of several distinct orders of individuals or castes which live together in populous, organised communities. In both there are, besides the males and females, a set of individuals of no fully developed sex, immensely more numerous than their brothers and sisters, whose task is to work and care for the young brood. In true ants this class of the community consists of undeveloped females, and when it comprises, as is the case in many species, individuals of different structure, the functions of these do not seem to be rigidly defined. The contrary happens in the Termites, and this, perhaps, shows that the organisation of their communities has reached a higher stage, the division of labor being more complete. The neuters in these wonderful insects are always divided into two classes—fighters and workers; both are blind, and each keeps to its own task, the one to build, make covered roads, nurse the young brood from the egg upwards, take care of the king and queen, who are the progenitors of the whole colony, and secure the exit of the males and females when they acquire wings and fly out to pair and disseminate the race; the other to defend the community against all comers. Ants and Termites are also widely different in their mode of growth, or, as it is called, metamorphosis. Ants, in their early stage, are footless grubs which, before they reach the adult state, pass through an intermediate quiescent stage (pupa) enclosed in a membrane. Termites, on the contrary, have a similar form when they emerge from the egg to that which they retain throughout life, the chief difference being the gradual acquisition of eyes and wings in the sexual individuals during the later stages of growth. Termites and true ants, in fact, belong to two widely dissimilar orders of insects, and the analogy between them is only a general one of habits. The mode of growth of Termites and the active condition of their younger stages (larva and pupa) make the constitution of their communities much more difficult of comprehension than that of ants; hence how many castes existed, and what sort of individuals they are composed of, if not males and females, have always been puzzles to naturalists in the absence of direct observation. What a strange spectacle is offered to us in the organisation of these insect communities! Nothing analogous occurs among the higher animals. Social instincts exist in many species of mammals and birds, where numerous individuals unite to build common habitations, as we see in the case of weaver-birds and beavers; but the principle of division of labor, the setting apart of classes of individuals for certain employments, occurs only in human societies in an advanced state of civilisation. In all the higher animals there are only two orders of individuals as far as bodily structure is concerned, namely, males and females. The wonderful part in the history of the Termites is, that not only is there a rigid division of labor, but nature has given to each class a structure of body adapting it to the kind of labor it has to perform. The males and females form a class apart; they do no kind of work, but in the course of growth acquire wings to enable them to issue forth and disseminate their kind. The workers and soldiers are wingless, and differ solely in the shape and armature of the head. This member in the laborers is smooth and rounded, the mouth being adapted for the working of the materials in building the hive; in the soldiers the head is of very large size, and is provided in almost every kind with special organs of offence or defence in the form of horny processes resembling pikes, tridents, and so forth. Some species do not possess these extraordinary projections, but have, in compensation, greatly lengthened jaws, which are shaped in some kinds as sickles, in others as sabres and saws. The course of human events in our day seems, unhappily, to make it more than ever necessary for the citizens of civilised and industrious communities to set apart a numerous armed class for the protection of the rest; in this nations only do what nature has of old done for the Termites. The soldier Termes, however, has not only the fighting instinct and function; he is constructed as a soldier, and carries his weapons not in his hand, but growing out of his body. Whenever a colony of Termites is disturbed, the workers are at first the only members of the community seen; these quickly disappear through the endless ramified galleries of which a Termitarium is composed, and soldiers make their appearance. The observations of Smeathman on the soldiers of a species inhabiting tropical Africa are often quoted in books on Natural History, and give a very good idea of their habits. I was always amused at the pugnacity displayed, when, in making a hole in the earthy cemented archway of their covered roads, a host of these little fellows mounted the breach to cover the retreat of the workers. The edges of the rupture bristled with their armed heads as the courageous warriors ranged themselves in compact line around them. They attacked fiercely any intruding object, and as fast as their front ranks were destroyed, others filled up their places. When the jaws closed in the flesh, they suffered themselves to be torn in pieces rather than loosen their hold. It might be said that this instinct is rather a cause of their ruin than a protection when a colony is attacked by the well-known enemy of Termites, the ant-bear; but it is the soldiers only which attach themselves to the long wormlike tongue of this animal, and the workers, on whom the prosperity of the young brood immediately depends, are left for the most part unharmed. I always found, on thrusting my finger into a mixed crowd of Termites, that the soldiers only fastened upon it.* Thus the fighting caste do in the end serve to protect the species by sacrificing themselves to its good. A family of Termites consists of workers as the majority, of soldiers, and of the king and queen. These are the constant occupants of a completed Termitarium. The royal couple are the father and mother of the colony, and are always kept together closely guarded by a detachment of workers in a large chamber in the very heart of the hive, surrounded by much stronger walls than the other cells. They are wingless, and both immensely larger than the workers and soldiers. The queen, when in her chamber, is always found in a paired condition, her abdomen enormously distended with eggs, which, as fast as they come forth, are conveyed by a relay of workers in their mouths from the royal chamber to the minor cells dispersed throughout the hive. The other members of a Termes family are the winged individuals. These make their appearance only at a certain time of the year, generally in the beginning of the rainy season. It has puzzled naturalists to make out the relationship between the winged Termites and the wingless king and queen. It has also generally been thought that the soldiers and workers are the larvæ of the others: an excusable mistake, seeing that they much resemble larvæ. I satisfied myself, after studying the habits of these insects daily for several months, that the winged Termites were males and females in about equal numbers, and that some of them, after shedding their wings and pairing, became kings and queens of new colonies; also, that the soldiers and workers were individuals which had arrived at their full growth without passing through the same stages as their fertile brothers and sisters. A Termitarium, although of different shape, size, texture of materials, and built in different situations, according to the species, is always composed of a vast number of chambers and irregular inter-communicating galleries, built up with particles of earth or separate matter, cemented together by the saliva of the insects. There is no visible mode of ingress or egress, the entrances being connected with covered roads, which are the sole means of communication with the outer world. The structures are prominent objects in all tropical countries. The very large hillocks at Santarem are the work of many distinct species, each of which uses materials differently compacted, and keeps to its own portion of the tumulus. One kind, *Termes arenarius*, on which these remarks are chiefly founded, makes little conical hillocks of friable structure, a foot or two in height, and is generally the sole occupier. Another kind (*Termes exiguus*) builds small dome-shaped papery edifices. Many species live on trees, their earthy nests, of all sizes, looking like ugly excrescences on the trunks and branches. Some are wholly subterranean, and others live under the bark, or in the interior of trees; it is these two latter kinds which get into houses and destroy furniture, books, and clothing. All hives do not contain a queen and her partner. Some are new constructions, and, when taken to pieces, show only a large number of workers occupied in bringing eggs from an old over-stocked Termitarium, with a small detachment of soldiers, evidently told off for their protection. A few weeks before the exodus of the winged males and females a complete Termitarium contains Termites of all castes, and in all stages of development. On close examination I found the young of each of the four orders of individuals crowded together, and apparently feeding in the same cells. The full-grown workers showed the greatest attention to the young larvæ, carrying them in their mouths along the galleries from one cell to another, but they took no notice of the full-grown ones. It was not possible to distinguish the larvæ of the four classes when extremely young, but at an advanced stage it was easy to see which were to become males and females, and which workers and soldiers. . . . . Thus I think I made out that the soldier and worker castes are, like the males and females, distinct from the egg; they are not made so by a difference of food or treatment during their earlier stages, and they never become winged insects. The workers and soldiers feed on decayed wood and other vegetable substances: I could not clearly ascertain what the young fed upon, but they are seen of all sizes, larvæ and pupæ, huddled together in the same cells, with their heads converging towards the bottom, and I thought I sometimes detected the workers discharging a liquid from their mouths into the cells. The growth of the young family is very rapid, and seems to be completed within the year: the greatest event of Termite life then takes place, namely, the coming of age of the winged males and females, and their exit from the hive. It is curious to watch a Termitarium when this exodus is taking place. The workers are set in the greatest activity, as if they were aware that the very existence of their species depended on the successful emigration and marriages of their brothers and sisters. They clear the way for their bulky but fragile bodies, and bite holes through the outer walls for their escape. The exodus is not completed in one day, but continues until all the males and females have emerged from their pupa integuments and flown away. It takes place on moist close evenings, or on cloudy mornings: they are much attracted by the lights in houses, and fly by myriads into chambers, filling the air with a loud rustling noise, and often falling in such numbers that they extinguish the lamps. Almost as soon as they touch ground they wriggle off their wings, to aid which operation there is a special provision in the structure of the organs, a seam running across near their roots, and dividing the horny nervures. To prove that this singular mutilation was voluntary, on the part of the insects, I repeatedly tried to detach the wings by force, but could never succeed while they were fresh, for they always tore out by the roots. Few escape the innumerable enemies which are on the alert at these times to devour them; ants, spiders, lizards, toads, bats, and goat-suckers. The waste of life is astonishing. The few that do survive pair and become the kings and queens of new colonies. I ascertained this by finding single pairs a few days after the exodus, which I always examined and proved to be males and females, established under a leaf, a clod of earth, or wandering about under the edges of new tumuli. The females are then not gravid. I once found a newly-married pair in a fresh cell tended by a few workers. The office of Termites in these hot countries is to hasten the decomposition of the woody and decaying parts of vegetation. In this they perform what in temperate latitudes is the task of other orders of insects. Many points in their natural history still remain obscure. We have seen that there are males and females, which grow, reach the adult winged state, and propagate their kind like all other insects. Unlike others, however, which are always, each in its own sphere, provided with the means of maintaining their own in the battle of life, these are helpless creatures, which, without external aid, would soon perish, entailing the extinction of their kind. The family to which they belong is therefore provided with other members, not males or females but individuals deprived of the sexual instincts, and so endowed in body and mind that they are adapted and impelled to devote their lives for the good of their species. But I have not explained how these neuter individuals, soldiers and workers, come to be distinct castes. This is still a knotty point, which I could do nothing to solve. Neuter bees and ants are known to be undeveloped females. I thought it a reasonable hypothesis, on account of the total absence of intermediate individuals connecting the two forms, that worker and soldier might be in a similar way female and male whose development had been in some way arrested. A French anatomist, however, M. Lespès, believes to have found by dissection imperfect males and females in each of the castes. The correctness of his observation is doubted by competent judges: if his conclusion be true, the biology of Termites is indeed a mystery. The different forms of which Lespès and Dr. Hagen speak, I could not find in the species observed by me. I found, however, a species whose soldier class did not differ at all, except in the fighting instinct, from the workers." (?)

THE BEE NATION.

CHAPTER XVII

Royalty.

WE have already seen that the Termites shun the light of day, and must, therefore, be reckoned among the decided "darkies." This is also shown to some extent in their State polity, which, as already said, otherwise much resembles the Ant Republic, but which approaches the monarchical idea by possessing a standing army and having generally only one queen. By this possession of a standing army the Termites' State is rendered more monarchical than the famous Bee polity, so often regarded as the prototype of a monarchy, or of the rule of an individual; the latter indeed, as a rule at least, has only one queen, but instead of a standing army, it carries out to the fullest extent the purely republican or democratic principle of universal national arm-bearing, in a fashion that leaves far behind it all human arrangements. But not in this alone, but in all its affairs, the bee State must be characterised as a monarchy with very democratic institutions. It may, indeed, be called a communistic or social-democratical monarchy—such as Napoleon III. for a time, while coquetting with the working-classes, appears to have had the notion of introducing in France. It may also be called an elective monarchy, for no direct hereditary line is followed, but the queen is in each case chosen by the workers, and selected or rejected as they please. The queen in return, relies wholly upon the workers, or the neuter working bees, of which there are from ten to sixty thousand in a hive, and which, by the possession of their terrible poisoned sting, unite in their own persons the positions of workers and soldiers; the privileged condition of the non-working, pleasure-loving males, or drones, is only suffered by the workers, as we shall presently see, just so far and for so long as their services are thought necessary.

On the other hand the monarchical principle is very

plainly manifested in the fact that the whole life of the hive revolves more or less round the queen; where she is wanting, dies, or is not succeeded by another, the hive falls into disorder, and in a longer or shorter time infallibly perishes. Single members of the hive, if they scatter, either die or become useless, lazy, vagabonds, and mischievous highwaymen. The monarchical principle of the bee nation is still more strikingly manifested in comparison with the other social insects, in that only one ruler or queen is permitted, and that where several accidentally come together the superfluous ones are either killed or are compelled to go out and found new colonies.

We therefore find that the bees, under all circumstances, have but a single queen, and that they obey implicitly the famous maxim of Homer, so often quoted in the interest of the political rule of one:

" Οὐκ αγαδὸν πολυκοἰρανίη, εις κοίρανος ἐστω."

(The rule of many is not good; let one be king!)

Nevertheless we sometimes find that an old and abdicated queen, no longer able to lay any fertilised eggs, is out of mercy suffered to remain for awhile in the hive near her successor, and receives some measure of the bread of charity. Pfarrer Calminius (No. 21 of the "Bee Journal," 1855) observed a case in which two queens lived peaceably and well-cared-for near each other on two tables hanging side by side. But these are rare exceptions. The workers generally sting the old useless queens unmercifully to death, or suffocate them by surrounding them closely on all sides. Sometimes they are merely driven out of the hive and perish hopelessly. It is, therefore, impossible to free the bees from the reproach of republican ingratitude, and in this matter—practical as is their behavior—they are decidedly behind men, who, when they get a new ruler, reckon it a point of honor to provide for the old and surviving one, and for all his cousins and relations into the bargain!

The more remarkable, in contrast, is the behavior of the bees to the real reigning and egg-laying queen, who is always treated with all imaginable love and care, and is constantly followed by a court of young bees which anticipate all her wishes and needs. Especially is built for her, or rather for her larvæ, a dwelling, or cradle, which is large and splendid compared with the small and narrow cells of the drones and workers; this is called the king's or queen's cell, or royal cradle, and is decorated with three-cornered stars, and requires a hundred times more wax for its erection than the ordinary cells, although the wax is of a specially costly substance, difficult to obtain, in the use of which great frugality is otherwise shown by the bees. The young bees divide the wax into thin transparent plates between the somites of their abdomens, and require for this much food as well as rest and warmth, so that they are only found in the interior of the nest and scarcely ever quit it.

A queen lives four years as a rule, and so long as she lives and is well all is in order in the state. But if any accident happen to her it is at once noticed by her people; they become restless, cease work, and make an uneasy noise which can be clearly heard from the interior of the leaderless hive. Attacks from without on the queen are heavily avenged by the bees, so that it is dangerous to take away the queen from a swarm, or to kill her. The much used name of "guide" has been given to the queen, because it was thought that she was a male, and acted as guide or lieutenant during swarming. On the same ground those whom the extraordinary attachment of the hive to the supposed ruler did not escape, called the old ones kings. From this point of view the Roman poet Virgil, in the 4th song of his famous poem on Agriculture, describes this attachment as follows:

> " Not e'en Egyptians in such reverence hold
> Their sovereigns; such respect has ne'er been felt
> By those in Crœsus' spacious realms who dwelt,
> Nor Scythians, nor those far tribes who live
> On Indus banks. When safe the queen, the hive
> Is all at peace: but if they lose their queen
> Their bond of union is broken; e'en
> Themselves they suck their honey-stores; destroy
> Themselves their comb's pierced fabric. Their employ
> Is by the queen arranged: her all regard;
> Surround with buzz continuous and guard
> In swarms; and oftentimes, when wars arise,
> Its queen upon its shoulders raising, dies
> The bee a noble death through countless wounds."*

---

* [Taken from Mr. Millington's metrical translation of the 4th Georgic. Virgil wrote "king" not "queen," but his translator has set the poet right!—TR.]

Good as may be the treatment by the working bees of their king, or rather their queen, that which they show towards the lazy and defenceless husbands, the males or so-called drones, is bad and even cruel. The queen-bee lives in a married state which is also often found among mankind, although rarer than its reverse, that known as polyandry. Her male harem is larger than almost any female harem among Orientals, for it consists of many—six to eight—hundred drones, which play a generally useless part in the bee state, for a single drone is enough to fertilise the queen, and they neither work, nor, lacking the terrible poisoned sting, can they protect and defend the State. They therefore thoroughly represent a hereditary peerage, which lets itself be served and fed by an industrious working class, without directly contributing anything to the good of the community; from May to August they lead an easy life, devoted to amusement, untouched by care or toil. If, indeed, they could foresee the woful fate which awaits them at the end of this period, their bliss would be less untroubled. Their great number—which, as has been said, far exceeds real necessity—would be a thoroughly incomprehensible and puzzling fact in the otherwise well-ordered bee State, if it were not to be regarded as a legacy from the formerly wild and uncultivated condition of the bees, in which each bee colony lived independently, and partly because of this, partly because of the many dangers threatening the drones on their flight, a very great number of these was requisite for the secure attainment of the object of their existence; to-day, when as a rule many hives stand close together, and the care and providence of men ward off dangers, so large a number of drones no longer seems necessary.

This mistake of nature, however, is corrected, or set right again by the prudent workers, which only put up with and feed their idle brethren so long as they consider them necessary for the impregnation of the queen. But in autumn, or late summer, when the wedding flight is over and when food is getting scarce, the famous massacre of the drones takes place, in which the male aristocracy of the state is offered up for the common good, without regard to close family ties between them and the workers. The latter in thousands surround the fat, lazy, defenceless creatures, drive them together into a heap, and either pierce them with their poisoned stings, or, after they have weakened them by hunger, throw them out of the hive, and they perish the next cool night from cold and hunger—so that in the autumn and late summer masses of dead drones are often found lying in front of the beehives. The drones' cells are then torn down, and the chance drones' eggs and pupæ found therein are thrown out; nothing is left alive which recalls idleness and sloth. Any drones which have escaped slaughter on the first day in their own hives, are sought out and massacred on the following. There can be no doubt as to the special ground of this murderous onslaught. The industrious working bees know that during the long winter season the drones, as unproductive consumers, can only hinder the life and well-being of the hive, without being of the smallest use to it, the queen already having been long fertilised. They therefore kill them, obeying the well-known maxim : " He that will not work, neither shall he eat." O shortsighted reason of the bees! Did ye know that among men those generally eat most and best who work least or not at all, ye would perhaps act more wisely!

That the massacre of the drones is not performed entirely from an instinctive impulse, but in full consciousness of the object to be gained, is proved by the circumstance that it is carried out the more completely and mercilessly the more fertile the queen shows herself to be. But in cases when this fertility is subject to serious doubt, or when the queen has been fertilised too late or not at all, and therefore only lays drones' eggs, or when the queen is barren, and new queens, to be fertilised later, have to be brought up from working-bee larvæ, then all or some of the drones are left alive, in the clear prevision that their services will be required later. In such hives living drones are often found all through the winter and even in the spring, and this is otherwise exceptional. The foresight and prudence of the workers is so great that in the first year of the foundation of a new colony they do not permit its queen, whose single impregnation suffices, as a rule, for many years, to lay drones' eggs, for they build no drones' cells—which owing to the larger bodies of the drones have to be larger and wider than those of the ordinary working-bees—and they either make the queen, who is able to lay one or other kind of eggs as she will, understand that she must only lay fertilised

P 2

eggs, or else make the laying of unfertilised or so-called drones' eggs impossible. The first idea is the more likely, for in case of need drones' eggs are placed in the smaller cells and these develop into drones. They are under these circumstances smaller than usual, but more room is made for them by the workers by the latter not roofing in the cells flat as for the ordinary broods, but giving them a raised roof. Such a brood is called a " humped brood," owing to the humped appearance of the cells. We see clearly here, as by so many other examples, that the bees are able to accommodate themselves to circumstances thoroughly and consecutively.

This wise calculation of consequences is further exemplified in that sometimes the massacre of the drones takes place before the time for swarming, as, for instance, when long-continued unfavorable weather succeeds a favorable beginning of spring, and makes the bees anxious for their own welfare. If, however, the weather breaks and work again becomes possible, so that the bees take courage anew, they then bring up new drones and prepare them in time for the swarming. This killing of drones is distinguished from the regular drone massacre by the fact that the bees then only kill the developed drones, and leave the drone larvæ, save when absolute hunger compels their destruction. Not less can it be regarded as a prudent calculation of circumstances when the bees of a hive, brought from our temperate climate to a more southern country where the time of collecting lasts longer, do not kill the drones in August, as usual, but at a later period suitable to the new conditions.

A bee State without drones is a real female State in the fullest sense of the words, for it contains only fertile females and females with rudimentary sexual organs. Even the presence of the drones alters this character very little or not at all, for, as will be shown, these only play a very secondary part, and the whole far-reaching intelligence of the bees, as of other social insects, must clearly for the most part be inherited from the mother's side. Amongst men also, according to the assertion of famous writers, remarkable mental gifts have been inherited more from the mother than from the father. In any case the *rôle* of the male element in the bee State is so subordinate to the female, that the highest ideal of female desire for emancipation appears here to be realised, and that our champions of emancipation should here find an unexpected support of their theory.

The poor drones, or males, find themselves completely in the power and under the dominion of their working sisters. But even the queen herself, great as is the love and honor usually shown to her, is not secure from the stings of her democratically-minded subjects if she do not fulfil her royal duties in every respect as expected and demanded of her. When, for instance, the time for swarming, the division of the colony, has arrived, the old queen sometimes decides only reluctantly to leave the dear home-hive, and make room for her younger rivals. She comes out, followed by a crowd of her dependents, but soon returns to the hive, still accompanied by them. But if this is repeated two or three times without the queen taking flight, the bees, angered at the repeated deception, fall upon her and kill her, either with their stings or by smothering her, or they pull her off the board so that she may perish outside the hive—a friendly proceeding termed "an exile," by bee-masters. Men are wont to be more lenient towards the faults or weaknesses of their kings, and the banishment of them by their rebellious subjects but seldom repeats itself in history, while, as a rule, when the matter is reversed, men are not so punctilious, and the banishment of rebellious and disloyal subjects is a valued and much used privilege of human rulers.

A late-impregnated queen, who lays more drones' eggs than others, is also not safe from her subjects, which insist on the greatest order and regularity being observed in their household. "A female shut up for thirty days in June was set free and came back impregnated. From the beginning of June until November she laid only male eggs, and continued the same proceeding in the following April. The working bees became wild and unsettled, and killed her in May" (Giebel, "Natural History of Animals," iv., p. 191). It has already been mentioned that older queens, which in consequence of exhaustion of their store of semen can lay no more fertilized eggs, share a similar fate.

When the weather is bad, so that the old queen is unable to swarm and found a new colony soon enough before the emergence of her young rivals, the bees will kill her or drive her by force out of the hive, unless, on the other hand, the royal brood has been killed on account of the impossibility of swarming. Her endeavors to render impossible the emergence of the young queens who threaten her rule, by hurrying to the cells in which the royal larvæ lie, tearing them out and killing them, are generally frustrated by the care of the working bees watching the brood, so that at length nothing remains to her but to leave her ungrateful subjects, accompanied by her dependents, and to found a new hive. Sometimes the working bees let the old queen carry out her murderous intentions, when, as the bee-masters say, there is no " inclination to swarm," or when the hive will not be overfull with the newly emerged bees, and there is therefore no need to divide. In other cases the killing of the young bees would render swarming impossible, and must therefore be prevented. The younger bees are most interested herein, as they long for " room for the flight of free souls," so they protect their future rulers or worship the rising sun, while the older bees hold rather to the old queen, and leave the hive with her and a few drones. This is not surprising, as the older bees are accustomed to their old and beloved queen and prefer her to a new and strange one. Yet they often take different sides, the exact peculiarities and reasons of which are until now as dim and mysterious as the manifold and often puzzling parties among men. Whether pride, vanity, self-interest and place-hunting have any part therein, as among men, is not yet revealed to the naturalist. But in any case this is certain, that jealousy and love of dominion, the burning desire to rule alone, is the motive which impels the queen, and leads her to deeds which exactly reproduce those which fill many sad pages of human rule with deeds that make the hair stand on end. Thence comes it that after the flight of the old queen concord is not re-established within the hive for some time, although the marvellously sensible working bees, as we shall immediately see, take no part in the strife, and even try to curb it by force. They feed the larvæ of the future queens in the royal cells in different ways, with the object of preventing their simultaneous emergence. They also retain an emerged queen in her cell until the time for swarming arrives. If the working bees fail in their object, and two or more young queens emerge simultaneously or close together, they then fight bitterly until one or other is victor. As already said, the working bees take no part in the duel of the pretenders to the crown, but look quietly on at the battle with crossed forelegs, and finally acclaim the conqueror and offer her their homage. As to the corpse of the vanquished, they content themselves with throwing it out of the hive. They behave throughout as prudent politicians, and that in two respects. First, in that they wholly accommodate themselves to the all-important result, and secondly that they let their rulers fight out their own battles and take no part in them themselves. Human rulers act very differently. When they want to fight out a question, it is the blood of their subject that has to be spilt, and however the quarrel may end the latter on both sides have the blows. *Quidquid delirant reges, plectuntur Achivi!* (Whatever the kings may rave about the blows fall on the people.)

But the young queens themselves behave no less prudently than their subjects in the battles, and seem to follow the well-know Falstaffean saying, that " Discretion is the better part of valor." Franz Huber ("New observations on Bees," published by G. Kleine, 1859) saw two young queens, which had come out of their cells almost simultaneously, fall angrily on each other, but at once loosen their grasp when they found that the use of their stings would, in the position in which they were, kill them both. A few minutes later the rivals again attacked each other, but the result of the meeting was the same as before. The onlooking working bees seemed very discontented with the cowardice of their rulers; they threw themselves in the path of the fugitives and tried to hold them. On a third attack one succeeded in approaching her rival unseen. She seized her at the root of the wings with her teeth, mounted on her body, and drove her sting without difficulty right through the abdomen. The vanquished doubled itself up, crawled helplessly forward, quickly lost its strength and soon died. Huber made similar observation several times; the working bees always tried to hinder the flight of the combatants, while they gave them plenty of room so long as they were moving towards each other, and during the battle itself formed a regular ring round them.

If the young queens, as is usual, emerge at different times, the one which has emerged first tries to inflict on her yet unborn rivals the same fate which the old queen tried to inflict on her; she tears open the royal cells, and smothers the inmates. The working bees only interfere with her so long as there are enough bees to form a new swarm. In this case she is compelled to leave the hive in the same way as was her predecessor, and in this way, during the course of a single summer, or only a few weeks, three, four, or more swarms are sent off one after the other. They become, naturally, continually weaker, and are called secondary swarms. When the hive has by this repeated swarming become sufficiently thinned, the working bees no longer tend and guard the young queens in the former fashion, but let them fight each other without hindrance, until only a single one remains. If a strange queen be placed in a hive already provided with a chief, she is at once stung or smothered to death by the working bees.

Sometimes it happens that in consequence of these fights and by often repeated swarming all the queens in a hive have disappeared, and this, as already said, means the necessary destruction of the hive if the want cannot be supplied. The bees become restless, cease working and scatter. The younger ones fly away, the elder remain in the hive to die. The bee-masters know that the queen bee is dead by the cessation of all life in and around the hive, and by hearing in the interior a dull or sad complaining noise. But, remarkable to say, these striking changes do not take place when at the time at which the hive has become queenless, there are either royal pupæ from which new queens will shortly emerge, or when there are in some of the cells of the hive working bee young, not more than three days old, for the bees know that they can by special treatment develop other new queens from such working bee eggs or larvæ, and indeed they manage this important bringing up business with the care and cleverness which distinguish all their actions. They first select the young working bee larvæ which are to be given the nursing necessary to change them into queens, and widen into royal cells the cells in which they are, by pulling down the contiguous partition walls. Then three neighboring cells are torn away and the larvæ and food in them carried away. A cylindrical enclosure is next built all round, whereby the rhomb-shaped floor is kept up, for by its destruction the cells and larvæ of the opposite side would suffer. The larva remains for three days in this cylindrical tube. But as it requires for its development a royal dwelling like a pyramid-shaped cell, the point of which must be directed downwards, the bees at the end of the third day pull away the cells lying beneath, sacrifice the larvæ contained therein, and use the wax thus obtained to build a second pyramid-like tube beneath the first.* The cell is lengthened as the larva grows; and it is continually and amply fed with the royal food, a specially prepared and very nourishing mixture of honey and pollen, set apart for the queens and their larvæ, and is continually watched and guarded in the most careful manner by bees which relieve each other. The stimulating influence of this peculiar kind of food so develops the sexual organs—which otherwise would have remained rudimentary—of the insects thus treated, that they finally become fertile queens or mothers, capable of maintaining and propagating the race. Sometimes also, so-called false or factitious queens are thrown off, which by eating the royal food have stimulated their sexual organs to further development, and which, without impregnation, lay a number of drones' eggs. They are therefore named drone-mothers, and the resulting hive drone-bearing.

While the bees thus spare no trouble and pains to remedy the fatal loss of their queen in the above fashion, they are yet not so short-sighted nor so driven by instinct as not at once to relinquish this slow and toilsome work if their loss be made good to them by an accident. Franz Huber, in getting rid of the young in a hive, sent in too much smoke, and many of the older bees, and among them the queen, escaped. Huber considered the hive as lost, when he found the queen the next day some distance off, in the midst of a knot of bees, and carried it back to the queenless hive. But how great was his surprise when he found that, in this short time, the queenless bees had begun and nearly finished three royal

---

* It may here be asked why the generally sensible bees do not simplify the matter by building a regular royal cell, into which they could carry the eggs or larvæ to be brought up. But leaving aside the fact that the carrying or dragging of the latter can be of no use, the building of real royal cells is more toilsome, slower, and especially costlier than the proceeding above described, which only requires the easily repaired loss of a few workers' cells and larvæ. So that here again the bees choose the shortest and simplest plan.

cells. He tore down two and left only the third. The following morning he saw, to his astonishment, that the bees, which now had their queen again and required no other, had carried away all the food from the royal cell—clearly to prevent the larva therein contained from developing into a queen!

In the same way bees which have a queen but which are given larvæ with royal food to take care of, pull them out of their cells and greedily eat up the food. But if they are without a queen, they change the cells into royal ones and nurse the larvæ into queens. The time which may elapse in such cases before the emergence of the queen-pupæ extends, or may extend, to fourteen days or more, and the bees, as has been said, remain perfectly quiet and depressed during this time, awaiting the coming event. The life of the hive here depends entirely on an idea, or on the picture in the brain of the little creatures of a future event that is outside the ordinary rule, and it is comprehensible neither by "instinct" nor by "inheritance," but only by a mental process.

Better and quicker than even the bees themselves, can men, as the bee-masters, come to the help of a chieftainless hive by introducing a new and full-grown queen. But this proceeding is attended with special difficulties, for the bees of a hive only suffer their own members in it, and chase away or kill bees of another colony, distinguishing them apparently by smell. The same fate would befall a strange queen, who is received on her arrival with an angry humming, were it not that man's ingenuity has invented the so-called queen-house. It is a little cage, woven out of fine wire-trellis, in which the queen is enclosed and introduced into a queenless hive. The trellis prevents the bees from killing the new comer at once, and gives them time to recognise her as a new queen, and to acknowledge or accustom themselves to her.

Such an event is admirably described by Major D. Schallich, of Ludwigsburg, in a letter kindly sent to the author, Nov. 17, 1875:—

"The minister of Laudenbach, in Verbach-Thal, is one of the most distinguished bee-masters in Wurtemburg. I saw at his place how he took the honey from the comb, and how the bees were not simple enough to build new cells, but without further trouble carried the honey into the cells built by their predecessors. If it were a question of instinct, they

should instinctively have built cells one time as well as
another. At the same minister's I saw a pretty little experiment. It is well known that the inhabitants of a hive
accept no strangers. The minister took a bee and put it
down among those which were on guard in front of another
hive. The latter at once fell on the involuntary intruder,
killed it, and threw it over. A hive happened to have lost
its queen, and a new ruler was to be given to it. But if a
strange queen were simply put down in front of the queenless hive, she would at once blindly and 'instinctively' be
put to death by the guards. The matter, therefore, had to
be managed craftily. If we had understood bee-language, a
careful speech, in which they would have been made aware
of the high honor conferred on them and their future duties,
would doubtless have been sufficient. The bees would have
had time enough to master their feelings of grief, and to
think of business. This time must be given them in another
way. With this object our bee-master had a pretty little
trap of finest wirework, a mouse-trap of the tiniest sort. In
this he put the queen with a small court, closed a little
opening with wax, and placed the apparatus in front of the
hive to which he desired to give a new sovereign. Naturally
the bees fell 'instinctively' on the little cage and tried to
kill the inmates, but the wire saved them. The murderers
pressed hotly round the cage; but suddenly they recognised
the majesty before them. Anger ceased; wonderingly and
dutifully they surrounded the queen. The news of the glad
event spread with lightning rapidity through the hive, and
a glad humming expressed the feeling of the bees. Crowds
left the hive and hastened to see the queen and to assure
her of their entire devotion. The popular vote was brief
and demanded unanimously the acceptation of the stranger
as queen. The guards, ashamed of their insulting reception,
stood aside; after careful investigation of the prison the
wax closing the opening was discovered and removed, and
the queen by the grace of God (or the minister—F. B.) was
led to the throne. Her suite also remained unhurt. I
scarcely think that the queen had previously taken any
solemn promises. She was met with absolute trust, for a
queen had never yet deceived her people! The event which
is the basis of this little story will be confirmed by the
minister of Laudenbach."

The artificial replacement of a dead or lost queen by another or by a stranger is the more easily effected the longer it happens after the loss, or the longer the bees have had to forget their former ruler. This will happen in from twenty-four to thirty hours. Huber gave a new queen to a hive that had been queenless for four-and-twenty hours. The bees nearest her touched her with their antennæ, felt her all over the body with their proboscides, offered her honey, fluttered their wings and formed a circle round the ruler. They then made room for others which acted in a similar way, and so the circle enlarged continually. They then all fluttered their wings and shook themselves without confusion or noise, as though they had made a very delightful discovery. When the queen set herself in motion, the circle opened, formed a hedge, and escorted her. When she arrived at the other side of the comb, where hitherto complete quiet had prevailed, the same reception was repeated. The workers busy at the royal cells stopped work, pulled out the royal larvæ, and then devoured the food stored up round them! From thenceforward the queen was recognised by all the people, and behaved herself as though perfectly at home. The bees, however, are exceedingly capricious creatures, and will one day willingly accept an introduced queen, and the next fall upon her with really diabolical fury, although she had been given to them twenty-four or forty-eight hours after their loss. In the attempt to introduce Italian bees into Germany, the Freiherr of Berlepsch always found that at least three queens out of four, in spite of the greatest care and prudence, were stung, smothered, maimed, or chased away, while in the succeeding summer his experience was exactly the opposite. In any case it seems that a queenless hive will receive willingly a strange queen only when the feeling or consciousness of queenlessness and helplessness has spread through the whole hive and is shared by each individual bee. So long as this is not the case, it cannot be surprising that a strange queen is treated and illused as such by the majority of the bees.

An interesting observation on this point is published by the Rev. George Kleine, of Luethorst, in his pamphlet on Italian bees and beekeeping (Berlin, 1855): "In order to give a German hive an Italian queen I take away," he says, "a fully-stocked hive and put in its place one with empty comb and a honey-table hung out, within which is the new queen, in a bee-cage and placed on a brood-comb. Many of the bees collecting honey, which had flown and were still flying from the hive taken away, return to the fresh hive, because it is on the stand to which they are accustomed and which is well known to them. But the moment they fly in, they at once become aware of the great change. They stop, do not know where they are, come out of the hole again without depositing their loads, fly off, look most carefully round the stand to assure themselves that they have made no mistake, and go in once more when convinced that they are at the right place. The same thing is repeated over and over again, until the bees at last bow to the incomprehensible and unavoidable, lay down their loads and set to work at those tasks made necessary by the new arrangements of the hive. But as all the newly arriving bees behave in similar fashion, the disturbance lasts until late in the evening, and the uncertainty and anxiety of the bees is so great that the bee-master cannot contemplate them without deep sympathy. Night, however, lightens their trouble; they learn to yield to the inevitable, and although the disturbance has not quite ceased by the following day yet the affairs of the new colony begin to get into order. By the third day everything is regular; the bees regard themselves as the rightful inmates of the new dwelling, this being shown by the fact that they no longer yield free passage to members of their former home which had flown away, but send them back as unauthorised intruders. The caged queen can be set free from her prison very soon, as a rule after four-and-twenty hours, for the feeling of the first-comers that they have no right to the new dwelling and have made some inexplicable mistake and cannot get right again, does not allow them to experience any hostility towards the imprisoned queen. They probably consider themselves as merely on sufferance, and feel that they should be grateful that no action is taken against them for their illegal entry, as generally happens in bee experience."

Who can deny that in this whole remarkable behavior of the bees a complete understanding of a changed situation is betrayed, and man himself could not show it better nor more clearly. The same comprehension and prudent foresight in an unusual situation are manifested in the following case. The wind threw down from the stand of a bee-master—a friend of the author's whose name will soon become known—a straw bee-hive, the inmates of which were surprised in full work, and no small disorder in the interior was the result. The owner repaired the hive, put the loose comb back in its place, and replaced it in such a manner that the wind could not again catch it, hoping that the accident would have no further results. But when he examined the hive a few days later, he found that the bees had left their old home in the lurch, and had tried to enter other hives, clearly because they could no longer trust the weather, and feared that the terrible accident might again befall them.

It is almost impossible to introduce a new queen into a chiefless hive with a factitious queen, because the bees are under the delusion that they already have one. The egg-laying worker will also readily smother the real queen that is given, because it believes itself to be one. The never-erring "instinct" does not therefore tell them of their fatal error, any more than it tells the leaf-cutting solitary humble-bee if it cuts its leaves too large or too small for the protection and covering of its eggs, or if it makes a mistake in the choice of the leaves. (Reimarus, *loc. cit.* 2nd ed. p. 181.) Nor can the hatred shewn as described by the queen-bee towards her royal relations, which drives her to murder them, be well described as the result of an instinctive tendency, evoked for the benefit of the community; since it is not clear why it should not be as good in the bee as in the ant State for several queens to live together, and since such an arrangement would rather promote than prejudice the welfare of the community. There are in fact some races, such as the Egyptian bees, which always have several queens. By the frequent driving out of old queens and the formation of new swarms a real advantage for the general spread of the race would be given. So that nothing can bear the blame save the desire of the queen, inherited from generation to generation, for personal rule, wherewith is conjoined the sad consciousness that the appearance of a rival in the old hive means her own abdication and the compulsion to seek a new home. In any case it ought not to be forgotten in judging these circumstances, that young queens are, as a rule, more fruitful than old ones, and that, therefore, by natural selection in the struggle for existence a tendency may have been gradually evolved in the bees themselves which would lead them to favor the change of sovereignty.

CHAPTER XVIII.

Swarming.

THE two greatest events in the bee State are the already-mentioned swarming and wedding flights, and the behavior of the bees in these important respects shows incontrovertibly that they are perfectly conscious both of the object to be attained and of the difficulties and dangers surrounding it. Swarming, or the foundation of a new colony, does not take place until "bee-takers," that is scouts or spies, have been sent out, which investigate the neighboring places and seek for the most suitable spot for the new colony. The extraordinary restlessness of those that remain behind clearly shows that they know what an important matter is in hand. The majority do not go about their ordinary work, but hang in thick beard-like masses outside the mouth of the hive. A loud clear hum is heard inside and outside the hive, which lasts through the whole night. According to F. Huber the queen is restless, lays her eggs irregularly or drops them, interchanges touches with the bees which come in her way, and climbs on their backs. No one offers her honey, as usual; she takes it herself out of the cells which lie in her way, and no homage is paid her. Those which are disturbed by her movements follow her, run idly over the comb and spread the commotion to the other parts of the hive. As soon as the queen has been all over the hive the restlessness is general. The workers no longer concern themselves with the young, those which return from collecting do not disburden themselves of the pollen, but run wildly about. The uproar raises the temperature of the hive so much that the wax begins to melt, and this helps to induce the undecided ones to swarm. The temperature rises to 27° or 32° R. [33¾° and 40° C, or 92¾° and 104° F.–TR.] a heat which is as a rule intolerable to bees. The bees crowded together at the mouth of the hive become so hot that those underneath, owing to the down-draught of the hot vapor, become bathed in sweat. Their wings become so damp that they are not in a condition to fly and cannot get far beyond the board.

When the restlessness and tumult have reached their height, the flight at last begins—providing that the weather is clear and sunny and that the scouts have brought good reports—after some of the bees, to omit no prudent precaution, have put in their honey-bags provisions enough for three or four days. The swarm darts up in the air with the swiftness of an arrow, but soon comes again to rest, for before the regular flight a kind of gathering and trial of strength takes place. A necessary condition is of course the presence of the queen. If this is wanting the swarm returns to the hive; otherwise the yet undecided bees gather more and more round the swarm, which, as a rule, settles and hangs round a bough of some tree standing near the mother hive. If two or more queens are present, as is often the case in vigorously swarming hives, the swarm divides, or if the swarm has settled the queens fight until only one remains.

The moment of the earlier gathering of the bees is the point which the bee-master must not miss if he wants to catch the young swarm and put it into a previously prepared hive—this only being possible also if the queen is present and is caught. If he does not succeed, the whole swarm, so soon as the gathering is complete, is off and away, to settle down on some place which seems to it suitable. As, however, several days are often lost over this gathering together, and as after arriving at their new home some time must elapse before they can bring in fresh food, it is evident how necessary and well-considered is the already mentioned carrying with them of provisions by the swarming bees. The queen is on the journey the object of the tenderest care, and some observers assure us that she is supported and carried by strong working bees. This is at least true as regards the older queens with somewhat worn out-wings, while the younger ones fly vigorously. Secondary swarms with young queens send out no scouts, as a rule, but fly at random through the air. They clearly lack the experience and prudence of the older bees.

The dwelling which the bee-catcher offers to his swarm is not accepted by it altogether carelessly and without further trouble, but is first carefully examined. If it does not meet

Q

with popular approval, as for instance if the interior be dirty or ill-smelling, or is too large or too small for the needs of the swarm, the bees leave it again to search for some other and often far distant place. If an empty hive is in the neighborhood which meets with the approval of the scouts, the swarm takes possession thereof.

M. de Fravière had the opportunity of observing the manner in which such an examination is carried on, and with what prudence and accuracy. He placed an empty beehive, made in a new style, in front of his house so that he could exactly watch from his own window what went on inside and out without disturbance to himself or to the bees. A single bee came and examined the building, flying all round it and touching it. It then let itself down on the board, and walked carefully and thoroughly over the interior, touching it continually with its antennæ so as to subject it on all sides to a thorough investigation. The result of its examination must have been satisfactory, for after it had gone away it returned accompanied by a crowd of some fifty friends which now together went through the same process as their guide. This new trial must also have had a good result, for soon a whole swarm came, evidently from a distant spot, and took possession. Still more remarkable is the behavior of the scouts when they take possession of a satisfactory hive or box for an imminent or approaching swarm. Although it is not yet inhabited they regard it as their property, watch it and guard it against stranger bees or other assailants, and busy themselves earnestly in the most careful cleansing of it, so far as this cleansing was impossible to the setter-up of the hive. Such a taking possession sometimes occurs eight days before the entrance of the swarm.

The swarming, which is so important for the maintenance and propagation of the race, can be very simply prevented by artificially widening and enlarging the hive. The nation thus has room enough to spread and to make new combs, and no longer experiences the need of sending out swarms. If it were instinct which impelled bees to swarm such a proceeding would be incredible. Bee-masters distinguish between artificial and natural swarming; the first is brought about by taking a number of bees forcibly from thickly populated hives, after they have been stupefied by smoke, and putting them into a prepared hive in company with a queen. The poor bees are so utterly terrified by this that they yield to the will and cunning of the bee-master. But the poetry of swarming, the so-called "swarm-down," is naturally brushed away by this rough proceeding.

"Fire, smoke, and noise," says an experienced bee-master in the *Bee Journal* (1862, Nordlingen, II., p. 380, etc.) "the accompaniments of this enforced transit, as though the whole home were in flames, have at last driven the hunted wanderers into the new and empty dwelling; they sit separately on the walls, glad no longer to hear the terrifying noise, no longer to feel the drumming at their heels. They recover; a dear leader is left to them, their queen is still among them, and they again crowd round her. Necessity drives them to work. Like burned-out folk they begin to raise a new home, and the new colony prospers if heaven sends fine days and food, or if zealous artificial nature helps with ample saucers of honey.

But what is going on in the old home? A portion of the population has not let itself be induced either by blows or by fire to quit the beloved hearth and leave the children sleeping in the cradles. At last shaking and drum-beating are past. They set to work to mend and support what has been injured; they feel themselves again, if only half, in their old place. But the hope of still possessing their queen is a delusion; their one and all is lacking. Soon they recognise their loss; one part of the house asks the other as to the delay of their leader. One after another rushes out of the door and seeks in anguish for the beloved. Hopeless and without certain news they turn back again, and now a universal wail sobs through the house; the walls stream with the sweat of agony. Even when the wailing is over, and the greater number have quieted down, a single voice recalls the common loss, and the weeping breaks out anew, until grief is extinguished by its own excess, and time pours its healing balsam on the bereaved, or until calculating man sends in a new queen, or a child of the lost sovereign after long space of time is brought up from the cradle to the throne. Outside stands the bee-master, listening to hear whether peace has not yet returned, and could weep in company with his sorrowing bees whose bereaved condition he has himself brought about."

Quite different to the eye and the heart of the friend of the bees is the natural or uncompelled swarming, although it depends more on chance and favorable or unfavorable circumstances. "There is fresh glad nature; a union of fortunate surroundings are necessary, good food, a sunny day, warm balmy air, if all are to succeed; there is the charm of fear and hope. .... The first drones have ventured out into the midday sun and whisper glad hope into the ear of the bee-master. There is a loud hum in the hive. .... In sultry nights some of the bees make their beds outside and scatter in the morning to go to their work. But on a fine warm morning not one flies away, as though something special were on foot. Some come out, surround their friends, bring them news from within and explain by fluttering movements the approaching flight. As soon as the bee-master approaches the garden he hears the swarming-music, more welcome than the best concert. The swarming bees fly zigzag, ever fresh crowds streaming out of the hive, one after another. Many in their haste fall to the ground, but pick themselves up again and join in the merry dance. There is silence at the mouth of the hive, but there is life in the air; clear rays of sunshine break through the little living cloud. Hither and thither sways the swarm; no place of rest has yet been chosen and the eye of the bee-master follows them ceaselessly. Now comes a sharp gust of wind, and blows down the swarming bees. Back on the old hive fall queen and people: the little heap already gathering on a branch falls off again. The hive is black—all is over. The fair hope has vanished; the beautiful swarming day has passed away in vain; a few pounds of honey have taken flight: sadly the owner wanders round his hives. .... Listen! a sound strikes his ear. *Tüt! tüt!* there it is again, and *Quak! quak!* in harmony.* He cannot hear enough of it. .... The day

---

* The *Tüten* and *Quaken* are caused by the just emerged or nearly emerging young queens, and serve as a sign of approaching swarming to the bee-master. As soon as a young queen is ready to emerge, she announces the fact by this sound of "*Quak!*" If no emerged queen answers with her "*Tüt*" she slips safely and quietly out, but wisely remains in her cell if she hears the jealous answering cry, and for so long as this is heard within the hive. This cry, arising from fear and jealousy, can be heard in each hive which has several young queens at

breaks. . . . . Ere the sun is high begins the prelude. Louder and louder peals the music; the swarms come out, and this time in earnest, for now on the war-dance depends throne and life. There, on the pear-tree, the crowd grows denser and denser. . . . . You may count twenty thousand, well armed for fight and work."

After the swarm has been caught in the swarm-net, it is placed in the prepared hive, carefully cleaned within. " Crowds rush out of the mouth of the hive ; but they soon stop, turn round, and beat their wings joyfully. They know that their queen is inside. The swarm is successful. The wildest and most unruly gather at the entrance and hum joyfully. A small portion of the army still remain on the tree; but they soon begin to move, and search up and down for the queen. They hear the hum, fly off, and on to the hive. The bough is soon deserted, and no corpse marks the field of battle. . . . . The most impetuous soon come out and scan the hive and the vicinity. Building begins inside; soon there is a rustling and sweeping, presently a flying off and carrying. Yet a few lively preliminaries, and all feel at home: the new state begins to flourish. But how is it in the old hive? Peace has entered there. Many caskets truly are empty, for the travellers could not be sent out into the unknown with empty hands. . . . Many a child, repenting the leaving, returns to the mother-home, and the younger sisters in the cradle are growing vigorously. The old mother may have gone, but hopeful daughters lie in the silence of the royal chamber; until they are ready to rule the household, all goes quietly on its way; the house is in perfect order. The domestic party-strife is over; each member goes peacefully to its work. The brilliant part, however, of all bee-keeping is the swarming time, for the poesy that is therein. In natural swarming is the poesy of bee-keeping."

It is always the old queen who heads the first, or primary

---

swarming time. The "*Tüt*" arises from the free, the "*Quak*" from the still enclosed queens, and is caused by the expiration of the air through the stigmata or air-holes found on either side of the body. That these noises can have no object save that of mutual communication is proved by the fact that a young queen emits no "*Quak*" after her emergence, while a free queen only calls when she hears the other, and remains dumb when she has no rival to fear. All this only, as said, concerns the young queens. The old queens do not emit this cry.

swarm, but not before she has laid eggs in the royal cells, out of which new queens develop after her departure. But the all-presaging and care-taking working-bees do not build these cells until they see the queen busy laying drones' eggs, for it is only after this laying that the body of the queen is thin and light enough to take flight. Also, as said above, the hive must be sufficiently populous, that is, the bees must have a surplus number large enough to make a swarm. If this be not the case, the building of royal cells is left undone, even though the queen be occupied in laying drones' eggs. Sometimes the excitement in a hive during swarming is so great and infectious that nearly all, or the greater part, of the bees therein take part in it, and the hive would seem deserted if the great number of the bees abroad and foraging, as well as the emergence of the young broods, did not quickly refill it.

Very interesting also is the way in which the bees guard the royal cells and the future queens contained therein from attacks of the already emerged and free queens during the time of swarming. All the cells are carefully watched by a number of workers, and as soon as the queen approaches them she is pinched, pulled and bitten until she retreats. This proceeding is often repeated during the day. As soon as the queen, standing still and leaning her breast against the comb, begins her accustomed song, it is as though the bees were electrified : they bend their heads and remain motionless. But as soon as the singing ceases the charm is over, and the renewed attempts of the queen to destroy the royal cells are repulsed in the same way. The enclosed and imprisoned queens, which do not venture or are not allowed to leave their cells so long as an emerged queen is calling, now and then push their proboscis through a little hole they have made in the wax, and let themselves be fed by the guards with honey. As soon as this is done they draw back the proboscis and the chink is waxed up again by the bees. The bees are able to distinguish the relative ages of each of the queens, and let them out of their cells exactly in order of age. It has already been said that they are able to voluntarily change or decide the time of their maturity.

When many swarms have been thrown off in this way, the number of the remaining bees is at length so small that they are no longer numerous enough to properly watch the royal cells. Several young queens then break out simultaneously, seek each other out and fight; the queen coming victorious out of the strife takes possession of the throne without opposition. The imprisonment of the young queens lasts longest in bad weather, which hinders the swarming.

Except at the swarming season the above described protection afforded by the bees to the royal cells is less careful, and indeed generally scarcely exists: this is clearly because the young queens are no longer needed to lead secondary swarms, and therefore their assassination in the larval or pupal state is seen without regret. It is less evident why, as F. Huber assures us, similar behavior is often seen in the primary swarm led by the old queen, and the latter is not, like the young queens, forbidden or prevented from approaching and even destroying the royal cells. Huber thinks that the respect felt by the bees to be due to a fruitful queen, once enfranchised by them, here comes into play, and is also clearly to be seen in other circumstances. Fortunately the old queens, for reasons yet unknown to us, do not make too frequent use of this prerogative, or the keeping and multiplication of bee colonies so long as the old queen lived would be impossible. Perhaps also it is again only the older bees which manifest their respect and leave their queen free, partly because of this, and partly from egoistic motives or conservative leanings, while the younger, like youth in general, follow revolutionary principles, and expecting reforms from the new sovereign, guard her from unnatural maternal attacks. All the proceedings here described, however, vary so much in different places and at different times, and the desire to swarm is so strong in the bees in some specially favorable places (as in the forest of Luneberg) that several swarming seasons take place instead of one, and swarm after swarm is thrown off, both from the old hives and the newly-formed colonies. With this increased desire to swarm are naturally united corresponding inclinations relative to protecting the queens.

The second great event in the bee state is the wedding flight, which in the secondary swarms takes place, as a rule, very soon after the foundation of the new colony, and often on the same day, and which is absolutely necessary for producing a fruitful posterity. The queen performs it when the weather is favorable—that is, warm, windless, and sunny, in company with her husband, the drones, and it lasts for two or three hours. The fertilising act itself is performed high up in the air, and always in the air: never, as with ants, in or on the dwelling. It has, therefore, not yet been accurately observed, and it seems as though a feeling of modesty prevented the queen from performing this act before the eyes of the crowd. Its results, on the other hand, are easily recognised by the condition of the sexual organs, in which, as a rule, the intromittent organ of the male with the adjacent sexual parts remain.

The working bees left behind know very well that the whole future existence of the colony hangs on the success of the wedding-flight, and the uncertainty makes them so restless that during this time no one can approach the hive without being attacked or stung. Perhaps they are afraid also, that any stranger approaching may place some difficulty in the way of the happy return of their queen. They therefore dance continually round the hive and its vicinity in narrow and wide circles, always keeping their heads turned towards the hive. They cannot well have any object in this save to impress as deeply as possible on their memory the whole aspect of the place, its appearance, scent, etc., so that they may be able to find it again with certainty on the now imminent flights. Perhaps also they wish to afford the returning queen a mark or guide for refinding the hive. Even when a number of single or of many hives together are on a bee-stand, every single bee is able to at once find its own hive, after they have impressed on their memory by carefully walking over the board, flying round the hive, etc., the appearance, smell, form of the mouth, and so on. Other insects also do the same. Bates several times saw the sand-wasps living by the Amazon *(Bembex ciliata, Monedula signata)*, which make holes for their young in the sandy banks, go round the place several times before flying away, in order to imprint the locality deeply on their memory so as to find it again. Their ability of refinding the place seems the more wonderful, as the hole itself becomes so filled up with the shifting of the sand that the human eye can see absolutely nothing there; the fact, therefore, speaks very highly for the wonderful acuteness of sense of these insects. We can observe that when our own native wasps are leaving a place which they wish to revisit, as for instance a half-eaten fruit, they try to print the impression of the place deeply in their minds by flying round it and touching it with their heads. The bees behave in a similar way in such cases. Dujardin put a dish of sugar in a niche in a wall a long way off a bee-stand. A single bee which discovered the treasure noted exactly the appearance of the locality by flying round the edges of the niche and pushing its head against it, and then flew away and returned after some little time with a number of its friends, which all fell upon the sugar.

If the queen does not return from her flight, the whole behavior of the bees manifests deep trouble and uneasiness, and they emit a peculiar plaintive cry, a sort of dull drawn-out wail. It is the same note as is heard, as above mentioned, from the interior of a queenless or bereaved hive. The greatest restlessness is seen inside and outside the hive. The bees go in and out continually without any decided object, as though they were seeking for their lost queen in all corners, and that even when it begins to get dark, or is dark, whereas under normal conditions darkness drives them all into the hive. With the described wail are mingled the hissing sharp notes to which angry or irritated bees give vent. On the other hand the wildest joy is shown when the queen returns prosperously, happy, and rich with promise. Full of delight they raise their extended hind-legs, fan the air quickly and unceasingly with their wings, and give a clear distinctive cry of joyful emotion, one which can be easily recognised as such. This is quite different to the hiss or wailing note of the queenless hive, and resembles the joyful, musical " swarming note " emitted by swarming bees. The note of calm content is deep, like the humming in the evening after a good day's foraging.

Just the same behavior as at the return of the queen, and the same note of pleasure, may be observed in bees when they reach the hive safe and sound after a flight, when threatened storm, tempest, or rain, or when some menacing danger is warded off by themselves or by foreign assistance. They are chiefly anxious during their flights about the weather, and even a threatening cloud in a clear sky will drive them back to the hive, while they will fly out fearlessly if the sky be completely overcast. It often happens that tired, exhausted bees fall down as though they were dead before reaching the shelter of home. If they are picked up and put into the hive the same note of joy is heard.

Sometimes the queen does not obtain the object of her wedding flight, but has to return with the marriage unconsummated. The flight is then repeated on the following day. If all has been successful and she returns impregnated, the workers, which at once recognise her condition, receive her with every sign of delight, and lead her within the hive, pressing round, caressing, brushing, cleaning her, and showing her every sign of respect. For an impregnated queen is, in the eyes of the bees, quite a different and a far more to be reverenced creature than is an unimpregnated or virgin. While they treat the latter with indifference, the former, as we have said, is the object of the tenderest care and attention. She is at once given a suite of from ten to twenty bees, which follow her everywhere, and supply all her wants.

Arrived within the hive, the queen after two or three days, when the first wax cells are ready, begins the chief and most important business of her life, egg-laying. In addition to her regular court, she is generally surrounded by a whole troop of workers which manifest their content by bending their heads before her, dancing up and down, licking and stroking her, and so on. Perhaps this crowding round her has also the object of keeping her warm on cool days, for the queen requires a tolerably high rate of temperature during egg-laying (not less than 10° to 12° R.)

The queen is able, during the rich foraging time of a populous hive, to lay some thousands of eggs daily, and actually lays daily some hundreds up to a thousand, so that the number of eggs laid during the summer may be twenty, thirty, or forty thousand. She prudently regulates her egg-laying by the amount of provisions collected in the hive, as well as by the number of the inhabitants, and the plentifulness of food. Thus in a weak hive and a cold season she only lays a few hundred eggs daily, but in populous hives and abundance of food from two to three thousands! The queen in her egg-laying is, like all bees, a great stickler for cleanliness; she feels with her sensitive flexible antennæ all over the interior of a cell in which she is going to deposit an egg, in order to see if it is empty, brushed, polished, properly swept, and generally fitted to receive one. If this investigation is satisfactory, as it generally is, she turns round and deposits an egg in it from her abdomen. If this experiment is made impossible to her by cutting off her antennæ, she lays no more eggs in the cells; she runs about uneasily over the comb, lets her eggs fall on the floor, where they dry up and perish, and prefers to stay in the combless parts of the hive, to which only a few specially attached bees follow her. At last, conscious of her complete helplessness and uselessness, she tries to leave the hive, and is not accompanied in her flight by a single working-bee.

The queen, as a rule, lays only one egg in a cell. If she accidently lets more fall into it the accompanying workers, according to the report of most observers, see to their proper division, but Huber denies this, and maintains that the workers could not grasp the eggs without injuring them, on account of their softness. From time to time the queen rests a little from her severe labor, creeping head first into a wide or drone's cell, and lying there motionless for some time. The position which she takes does not permit the accompanying bees to pay her the attention above described. None the less, they do not neglect, even under these circumstances, to form a circle round her, and to lick the portion of her abdomen left outside.

The queen apparently has it in her power to lay either drones' eggs or working-bees' eggs, accordingly as she fertilises or not the eggs passing through her oviducts with the male semen, by pressing the laden spermatophores by aid of voluntary muscles. Fertilised eggs produce working-bees or queens, unfertilised, drones. The latter are laid in the larger or drone cells, the former in the smaller or working-bee cells. The most various explanations, mechanical and accidental, have been given of the peculiar behavior of the queen, and of the question why she should in the one case fertilise the eggs and not in the other, but none of them have been proved to be true. It seems far more probable that the queen has a clear understanding of the object of her work, and lays drones' eggs or fertilised ones as they are required by circumstances. The bee-masters know very well that a queen which is brought with a young swarm into a hive which has only drones' cells built in it will rather drop her eggs than deposit them in the drones' cells, for drones are unnecessary in the first year and are only a loss to the colony, and working bees' eggs will not develop in drones' cells. The queen crawls to the sides where no drones are wanted, across the drones' cells without laying eggs therein. If she had no knowledge of the different objects of her egg-laying and was only impelled thereto by an instinct to general egg-laying mechanically controlled, such behavior would be incredible, and it would be a matter of indifference to the queen in what kind of cells she laid her eggs. It is also known that an injury to the last abdominal nerve ganglion makes it impossible for the queen to lay other than drones' eggs, for she can consequently no longer voluntarily act on the spermatophore. If, as has been mentioned, the cause is the mechanical pressure of the narrow working-bees' cells on the spermatophore, the queen thus injured would be able to lay male and female eggs as before. There can therefore be no doubt that the queen is able to voluntarily decide the sex of the laid eggs, and that she fertilises them or not with a thorough foreknowledge in each case of the task before her, just in the same way as she regulates the number of the eggs she lays according to circumstances.*

This seems the more probable when we learn that the queens seem to know exactly when they have discharged their queenly duties, and in such a case are seized with a presentiment of their appproaching end. The wonderful observation has been made that a queen who, through age or some other weakening circumstance, becomes conscious of her exhaustion, and has communicated this consciousness to her people, provides, in common with them, for the safe succession to the throne, and as soon as this is done gives back the throne and sceptre into the hands of the people, that is, either voluntarily leaves the hive in order to die outside, or is killed by the bees and thrown out of the hive.

Perhaps it is a feeling similar to this wonderful perception which, as already said, withholds old queens from making

---

* More exact details on this point and on the different explanations offered of it in an appendix to an essay of the famous American beemaster, Charles Dadant, will be found in an essay of the author's, "The German School in opposition to the American" in the "*Æsterreichischen Bienenzeitung,*" December, 1879. (Published by Rudolf Mayerhöffer.

use of their prerogative against youthful rivals to the same extent as the young queens sometimes do, and always wish to do, if the people did not interfere for protection.

CHAPTER XIX.

Domestic Work.

WITH the egg-laying of the queen begins the special work of the hive in connection with the maintenance and propagation of the family. The principle of division of labor, which we have already seen carried to so great an extent among ants, is here also fully developed.

The whole activity of the hive may be divided into two parts, domestic and abroad, the domestic being, as a rule, performed by the younger, and the work abroad by the older bees. This is very easy to establish, for the younger bees are readily distinguished from the older by color and outside appearance, and are recognisable especially by a fine white down and unsoiled wings. The easier domestic work, requiring less exertion, falls quite naturally to the share of the weaker bees, while the harder and dangerous work outside belongs to the older and stronger. Yet there are considerable grounds for the opinion that a number of older bees remain in the hive to give the younger ones instruction and guidance in their tasks. The young bees themselves, on emerging from the larval or pupal state, are not thoroughly fit and ready, and endowed with the capabilities of their race, as instinct-mongers imagine, but are so weak on the first day after their emergence that they cannot even fly. They require at least from twenty-four to thirty hours before their full strength and capacities are developed. The case of the queens would be similar, were it not that their imprisonment is, as a rule, prolonged over the time of change, so that during this interval they can become fully developed.

It cannot be doubted that, as said above, the young bees receive instruction and guidance from the elder, and this the more as the work within the hive is of a very various and complicated sort, and it cannot be imagined that each individual bee is born with the instinct for a special kind of work. Regard for the maintenance of the colony requires that each individual bee, in case of need, shall be able to undertake each kind of work, and this can be ascertained by direct observation. Since the queen only fulfils her maternal duties so far as the laying of the eggs is concerned, and troubles herself no further as to the fate of her children—she, indeed, having no time to do so—the whole care of the offspring falls at once on the shoulders of the bees of the hive, and they practically act as nurses as well as attendants. They arrange the cells, or the cradles for the reception of the eggs, clean them, and build new ones when wanted. They prepare the wax for building the cells, and also the so-called bee-bread, which is made out of honey, pollen, and water, and with this are fed not only the very hungry grubs and larvæ, but also the queen ; the drones are left to feed themselves from the store-chambers, the working-bees regarding them with a certain contempt or indifference. The queen, during egg-laying, requires a very large amount of nourishment, on account of which ten or twelve feeders are constantly busied with her. The young bees also, after they have left the cradle, require feeding for some days, until they can eat by themselves, and must also, like human babies, be washed and cleansed from the soiling consequent on birth. There is also a great difference between the food of the drone and working-bee and the royal larvæ ; the latter, during their whole life and until their emergence, are fed with bee-bread, whereas the former only receive this at the beginning, and are only fed with pollen and honey during the later days of larval existence. As the internal sexual organs develop during this period, it is clear why so great a difference is found in the further development of these organs in queens and working-bees. The bee-bread is nothing more than chyme made externally, the digestive process being half performed in the interior of the bees which feed them. It has, however, become more concentrated than chyme by evaporation in the cells. Undigested grains of pollen, which can be easily recognised in the small intestine of the working-bee larvæ in the last days of larval life, are not to be found in bee-bread; under the microscope it is an amorphous tenacious substance, with countless fine granules of fatty appearance.

The attendants set apart for the young brood have further

to cover the cells with a waxen roof when a larva begins to spin its cocoon, and later to reopen it to free the imago. When this is done, the soft silky cocoon is either pulled out or else firmly fastened like a tapestry to the interior of the cell; the cell itself is carefully cleaned and smoothed, so that it may, according to circumstances, serve either for a new egg or for honey. According to Huber the working-bees, as a rule, remain three days in the egg and five days as larvæ. After this lapse of time the bees close their cells with a covering of wax, while the larvæ begin to spin their silken shirts, taking thirty-six hours to complete their task. Three days later they change into a so-called nympha, and remain for seven or eight days in this condition: on the twentieth day they reach the imago condition, reckoning from the day on which the egg was laid. Their further development, if there is such, proceeds very rapidly, and is completed in a few days. The development of the queen, owing to her better food, is rather more rapid; that of the drones somewhat slower.

Just as with the nursery cells, the house bees have to close with wax the provision cells, or store-rooms, when they are filled with honey, so as to prevent it from running out. The more the cells are filled with honey, the more carefully do the bees prevent any outflow by gradually pulling the wax covering from the edges of the cell over the opening, and closing it in the middle as soon as the cell is completely filled. As these stores serve for food during the winter season, it is strictly forbidden to open them, and only in cases of extreme need, when honey is nowhere else to be found, are the wax coverings raised. During the foraging season, on the contrary, when food enough comes in from without, they are never opened. Other cells, always standing open, serve the bees for daily use, but no bees take more from them than is necessary for the satisfaction of their immediate need. The intemperance so customary among men as regards eating and drinking, which subserves excessive enjoyment and not necessity, is normally unknown to this exemplary creature, although, as we shall find later, the vices of gluttony and drunkenness lead many of them to destruction under special circumstances.

The preparation of the honey itself is entrusted to the house-bees, while the foraging bees only collect outside, and carrying off the nectar from the flowers in their honey-bags, put it in the lowest cells so as to fly off again as quickly as possible and collect fresh stores. The nectar is taken by the house bees, turned into honey in the interior of their bodies, and laid by in their upper cells. If the foraging is very successful so that these cells do not suffice for the reception of the wealth brought in, the wise little insects manage by extending the cells assigned to the honey to four or six times their original length before closing them. These long honey-filled cells, or rather tubes, then thoroughly deserve the name of store-rooms. These chambers have, however, the disadvantage that they seriously narrow the free space necessary for communication between the various combs. But as soon as spring comes, the bees have nothing more urgent to do than to bring these long and then empty cells back to their normal length, and so to again have the proper space between the combs. They also put the combs with the thus lengthened cells originally very far apart, and build between them, when by emptying them their length is become greater than is wanted, one-sided combs, or combs with only one row of cells.

Special attention is paid to the pollen brought to the hive, each sort—of which there are often from six to ten—being brought into special cells. These different kinds apparently serve for making different, and especially finer and more stimulating, sorts of food. The best and finest food is given to the grubs or larvæ from which are to be developed the future queens, both those in the ordinary royal cells and also those working-bee larvæ which in case of the failure of real queens shall be brought up as supplementary.

The constant cleaning and brushing of the queen, and of the working-bees returned from foraging, also takes much time and trouble, as well as the keeping clean of the hive itself. Every kind of dirt—rubbish, dead bees, wax coverings from opened honey cells, and everything not belonging to the hive—is carried away from it. Further, the bees try to make the interior of the hive as smooth, neat, warm, and secure as possible by the use of the so-called *propolis*, also called stopping wax and resin wax; with this they carefully stop up all the clefts and crannies, and thereby specially try to guard themselves against the attacks of the terrible wax moth. The interior of the individual cells is also smeared over with this substance, to make them firmer and more resistant; it is not like other wax, manufactured by the bees, but is collected from the resinous parts of trees and brought to the hive. F. Huber saw bees scraping off the propolis with their jaws from the legs of their comrades returning from outside, and hasten with it to the scarcely completed cells.* The store was here divided among a number of workers, which at once busied themselves with the suitable use of it. They first smoothed and planed the inner surface of the cells with their mandibles, freeing it from every unevenness. One of the workers then approached the heap of propolis near them, pulled a little thread out of the resinous mass with its jaws, tore it off by suddenly throwing back its head, grasped it with the claws of its fore-feet, and turned back with it into the cell which it had prepared. Without more ado it laid the thread between the two walls which it had smoothed, and on the floor of the corner made by them with each other. The thread proving too long, it bit a piece off. Placing it carefully with its fore-feet between the walls, it pressed it firmly with its mandibles into the corner which it wished to cover. As it was now seen that the band was too broad and heavy it was gnawed, and pieces carried away until it fitted. When the task was complete, the observer marvelled at the exactitude with which the little ribbon was fitted between the two walls of the cell. But the worker waited no longer over it, but turned to another part of the cell to treat the propolis remaining over in like fashion. Other bees completed the task which this one had begun, so that soon all the walls and openings of the cells were framed with bands of propolis.

By this process the wax cells, which are very soft and fragile when they come from their makers' hands, are clearly rendered more firm, and this is done without using too much

---

* The expression "saw" may perhaps surprise those who know that Franz Huber, the famous historian of the bees (born in Geneva, 1750, d. 1831), and the father of Peter Huber, the distinguished observer of the ants, became blind in youth owing to severe study. None the less he so successfully followed his remarkable study of bees by aid of his wife, a devoted companion (Franz Burnens), his son, and a few other friends, that his work, which was published in 1794, is to-day the chief source of information for all who employ themselves deeply with bee-life—although much contained therein needs correction.

of the wax, which is both costly and difficult to make. The fastening of the combs to the walls of the hive is also generally done by help of propolis, or of a mixture of propolis and wax. In the same way combs over-weighted with honey and threatening to break from their upper attachments, are so well supported and fastened with strong bands of propolis above and at the sides, that the peril is smoothed away. F. Huber saw such a comb which had really fallen down, but had kept in a proper position with respect to the other combs, at a time when there was not sufficient store of wax to fill up the empty space with new cells, fastened by such bands to the nearest comb on one side, and on the other to the neighboring glass wall of the hive. They also took warning from this unpleasant experience, and fastened the remaining combs to each other in similar fashion, and strengthened their upper points of attachment with old wax —clearly to ward off from these a similar misfortune. "I admit," says the generally instinct-believing Huber in relating this, "that I was unable to avoid a feeling of astonishment in presence of a fact from which the purest reason seemed to shine out."

A similar and even yet more striking observation is published by Dr. Brown in his work on bees (quoted by Watson, "The Reasoning Power in Animals," 1867, p. 448). A too heavily weighted comb had fallen down in the middle of the hive, and pressed against its neighbor, at the same time blocking up the passage. This accident caused great excitement in the colony, and brought about the following results. The bees first bound the two combs together with horizontal cross-pieces, and then gnawed away sufficient wax and honey to leave the passage free. The fallen comb and its neighbor were then fastened to the window with propolis, and the original cross-pieces taken away again, as being no longer necessary! The whole operation took about ten days. Men under similar circumstances could not have acted more prudently. The bee-master has many opportunities of remarking similar reasonable actions during the building in the interior of the hive. Dr. Dzierzon, of Carlsmarkt, the very experienced keeper of bees, and the discoverer of their parthenogenesis, who merits so many thanks for the exhibition of the inner relations of the beehive by the introduction of movable comb-boxes, thought himself justified in saying in an essay on the utility of bee-keeping :—" The cleverness of the bees in repairing perfectly injuries to their cells and combs, in supporting on pillars pieces of their building accidentally knocked down by a hasty push, in fastening them with rivets and bringing everything again into proper unity, making hanging bridges, chains and ladders, compels our astonishment."

The propolis is lastly used for an air-tight cover over creatures which have invaded the hive and which the bees have killed, but which are too large for them to carry away, such as mice, slugs, moths, etc., and in this fashion are prevented the injurious results to the hive which would follow the putrefaction of the corpses. For impure air within the hive is that which the bees must above all things fear and avoid, for with the pressure together of so many individuals in a comparatively small space, it would not only be directly harmful to individual bees, but would produce among them dangerous diseases. They therefore also never void their excrements within, but always outside the hive. While this is very easy to do in summer, it is, on the contrȧry, very difficult in the winter, when the bees sit close together and generally motionless in the upper part of the hive, and when, from impure air and foul evaporations, as well as from bad and insufficient food, dysentery-like diseases break out among them and often carry off the whole community in a brief space of time. In such cases they utilise the first fine day to relieve themselves, and in the spring they take a long general cleansing flight. But they also know how to take advantage of special circumstances so as to perform the process of purification in the way least harmful to the hive. Herr Heinrich Lehr, of Darmstadt, a bee-keeping friend of the author, has sent the following communication: During an epidemic of dysentery in winter, from which most of his hives suffered (as the bees were no longer able to retain their excrements) one hive suffered less than the others. Exact investigation showed that this hive was soiled all over at the back with the excrement of the bees, and that the inmates had here made a kind of drain. On this spot a little opening had been made by the falling off of the covering clay, which led directly to the upper part of the hive, where the bees were accustomed to sit together during the winter. This excellent opportunity, whereby they could reach in the shortest way an otherwise difficult object and one rendered complicated by circumstances, did not escape them.

The already often-mentioned love of cleanliness is a main characteristic of bees, both in reference to their dwelling and their persons. On entering a new dwelling the first thing they do is to cleanse it in the most careful manner from dust, dirt, wood-shavings, straws, etc. During the winter their body generally becomes covered with yellow-brown grease, which hinders their movements and injures their health. On the first fine day in spring, therefore, they first clean themselves, partly with their own hands, partly, in places which they cannot reach, by the help of their companions, and then, as already said, they cleanse and brush the interior of the hive with wonderful care, throw out old and hardened pollen, carry out mould and dead bees, etc., etc. That they also bury the latter appears from an observation quoted by Watson (*loc. cit.* p. 453) from the *Glasgow Herald* ("Notes and Queries," 3rd ser., vol. iii., p. 314). The correspondent writes: "While I was walking with a friend in a garden near Falkirk, we noticed two bees, coming out of a beehive, which were carrying between them the corpse of a dead comrade and flew away with it to about ten yards distance. We followed them and saw them search for a suitable hole at the side of a sandy road, carefully push in the dead body, head foremost, and finally place above it two small stones. They then watched for about a minute before they flew away."

The correspondent goes on to say that he had never until then had the opportunity of observing the burial of a bee, but that he had seen a wasp, which had invaded a beehive, after it had been killed, dragged out by the bees and laid down on the other side of a little brick wall, after they had flown over it with the dead body. It is also a very common observation that the bodies of dead bees are not left in the vicinity of the hive, but are carried away to a certain distance.

Very interesting, and closely connected with this characteristic of cleanliness, is the conduct of the so-called ventilating-bees, which have to take care that in summer or hot weather the air necessary for the respiration of the bees in the interior of the hive is renewed, and the too high temperature cooled down. This latter precaution is necessary, not only on account of the bees working within the hive, to whom, as already said, a temperature risen beyond a certain point would be intolerable, but also to guard against the melting or softening of the wax. The bees charged with the care of the ventilation divide themselves into rows and stages in regular order through all parts of the hive, and by swift fanning of their wings send little currents of air in such fashion that a powerful stream or change of air passes through all parts of the hive. Other bees stand at the mouth of the hive, which fan in the same way and considerably accelerate the wind from within. The current of air thus caused is so strong that little bits of paper hung in front of the mouth are rapidly moved, and that according to F. Huber, a lighted match is extinguished. The wind can be distinctly felt if the hand be held in front.

The motion of the wings of the ventilating bees is so rapid that it is scarcely perceptible, and Huber saw some bees working their wings in this way for five and twenty minutes. When they are tired they are relieved by others. According to Jesse the bees in very hot weather, in spite of all their efforts, are unable to sufficiently lower the temperature, and prevent the melting of some of the wax; they then get into a condition of great excitement, and it is dangerous to approach them. In such a case they also try to mend matters by a number leaving the hive and settling in large masses on its surface, so as to protect it as much as possible from the scorching rays of the sun.

Although the described plan of ventilation is remarkable enough in itself, it is yet more remarkable in that it is clearly only the result of bee-keeping and is evoked by this misfortune. For there could be no need of such ventilation for bees in a state of nature, whose dwellings in hollow trees and clefts of rocks leave nothing to be desired as to roominess and airiness, while in the narrow artificial hive this need at once comes out strongly. In fact the fanning of the bees almost entirely ceased when Huber brought them into large hives five feet high, in which there was plenty of air. It follows therefore that the fanning and ventilating can have absolutely nothing to do with an inborn tendency or instinct, but have been gradually evoked by necessity, thought, and experience.

One of the most important duties of the bees employed

within the hive is the building and arranging of the wax cells, to which only cursory allusion has as yet been made: these serve partly as cradles for the brood, partly for the reception and preservation of the collected provisions. The main principle here is, as many cells as possible with the greatest possible economy in wax, space, and labor—and this is realised by the bees in a manner that deserves our deepest admiration, and has indeed awakened the surprise of all observers. For as bees possess no mathematical and geometrical knowledge, and have also no need thereof, they have by practice, experience, and inheritance, aided by the principle of natural selection, gradually reached that kind and fashion of building which seems most fitted to their purpose. Each single comb or layer consists of two rows or divisions of cells arranged closely side by side; each cell with its six sides and its pyramidal floor is so arranged with respect to its fellows that the floors of one set make the roofs of those opposite them in the comb, and the sides serve as the boundaries of the neighboring cells; thus each wall, both of sides and floor, serves for two cells. But as this double duty imposed on the already thin walls, might easily give rise to the danger of breaking, the prudent insects make a suitable strengthening by surrounding the open edges of each cell with a thickening of wax, just in the same way that tinkers put a rim of tin round the edges of tinned ironware to make them stronger and less liable to bend.

It is very difficult to watch the bees during the absolute building of the cells. They are so eager to help each other, and press so closely together, constantly relieving each other, that there is seldom any opportunity of watching a single operation separately. But it can be seen that their mandibles are the chief instruments with which they spread out and smooth the wax. While some bring the six-sided cells to their normal length, others are busy laying the foundations of new ones: when the first hexagonal cells are shaped the others seem to take the same shape of their own accord. The wax left over in first forming them is carefully scratched off with the teeth, made into a little ball about the size of a pin's head and used elsewhere.

All the cells have not the same shape, as would be the case if the bees in building worked according to a perfectly instinctive and unchangeable plan. There are very manifold changes and irregularities. Almost in every comb irregular and unfinished cells are to be found, especially where the several divisions of a comb come together. The small architects do not begin their comb from a single centre, but begin building from many different points, so as to progress as rapidly as possible, and so that the greatest number may work simultaneously; they therefore build from above downwards, in the shape of flat truncated cones or hanging pyramids, and these several portions are afterwards united together during the winter building. At these lines of junction it is impossible to avoid irregular cells between the pressed together or unnaturally lengthened ones. The same is true more or less of the passage cells, which are made to unite the large cells of the so-called drone-wax with the smaller ones of the working-bees, and which are generally placed in two or three rows. The cells also which they usually build from the combs to the glass walls of their hives, in order to hold them up, show somewhat irregular forms. Finally, in places where special conditions of the situation do not otherwise permit, it may be observed that the bees, far from clinging obstinately to their plan, very well understand how to accommodate themselves to circumstances not only in cell-building but also in making their combs. F. Huber tried to mislead their instinct, or rather to put to the proof their reason and cleverness in every possible way, but they always emerged triumphant from the ordeal. For instance, he put bees in a hive the floor and roof of which were made of glass, that is of a body which the bees use very unwillingly for the attachment of their combs, on account of its smoothness. Thus the possibility of building as usual from above downwards, and also from below upwards was taken away from them; they had no point of support save the perpendicular walls of their dwelling. They thereupon built on one of these walls a regular stratum of cells, from which, building sideways, they tried to carry the comb to the opposite side of the hive. To prevent this Huber covered that side also with glass. But what way out of the difficulty was found by the clever insects? Instead of building further in the projected direction, they bent the comb round at the extreme point, and carried it at a right angle towards one of the inner sides of the hive which was not covered with glass, and there fastened it. The form and dimensions

·of the cells must necessarily have been altered thereby, and the arrangement of their work at the angle must have been quite different from usual. They made the cells of the convex side so much broader than those of the concave that they had a diameter two or three times as great, and yet they managed to join them properly with the others. They also did not wait to bend the comb until they came to the glass itself, but recognised the difficulty beforehand.

Huber further says that the bees can build their hexagonal cells on glass or wood instead of on a floor of wax, the floors of the cells being then necessarily flat instead of pyramidal. These cells with flat floors are less regular than the ordinary ones. Many of the cell edges are not angular, and the dimensions are often not exact. Notwithstanding, a more or less definite hexagonal shape can be recognised, even in those which depart most from the regular form.

The lower free end of a completed comb is also always finished off with a thicker border of wax with irregular cell-commencements. If the bees want to build on to this they first scrape off the rim of wax and the irregular cells, and then begin to carry the comb further. Pieces of comb hung like so-called screens are also ingeniously used by the bees for further building, but only after they have removed the injured rows of cells of the divided walls. But if such pieces are soiled, or seem to them to be otherwise unsuitable, they are torn down and a quite new comb is built. In the same way they make no difficulty, when they want to breed drones, about pulling down the working-bees' cells and making drones' cells instead. Thus we find everywhere a perfectly clear understanding of the condition of affairs, and an equally clear and definite adaptation of action according to circumstances.

These and many other observations show, as Huber says, " how ductile is the instinct of bees, and how readily it adapts itself to the place, the circumstance, and the needs of the community."

" Man must really be more stupid than a beast," says E. Menault (" L'Intelligence des Animaux," Paris, 1872) after describing similar proceedings of bees, especially as regards economy of space, " if he does not recognise in this conduct calculation, comparison, reflection, and reason."

Bees, like ants and like men, are ever subject to error, and often do things badly which afterwards have to be set right. Huber saw a worker building with the wax at her disposal on that which its comrades had already put together. But it was not arranged in the same way, and made a corner with the first. "Another bee perceived it, pulled down the bad work before our eyes, and gave it to the first in the requisite order so that it might exactly follow the original direction." Other observers, as for instance Darwin, have seen similar cases. "It was really curious," says Darwin ("Origin of Species," p 232), "to note in cases of difficulty, as when two pieces of comb met at an angle, how often the bees would pull down and rebuild in different ways the same cell, sometimes recurring to a shape which they had at first rejected."

CHAPTER XX.

ACTIVITY ABROAD.

LESS complex, but heavier than the work within the hive is that of the outside or foreign department. It consists almost entirely of the important business of collecting food for the young and for the hive, as well as the provisions necessary for the support of the community during the long winter season; the honey or nectar from the flowers is carried in a crop-like extension of the œsophagus, the pollen heaped in a shovel-like hollowed-out basin in the hind-legs, and brought home in the form of round pellets. The bees are sometimes seen so covered with pollen as to be scarcely recognisable. With wonderful rapidity they pull the pollen out of the flowers with their fore-legs, divide it with their middle-legs, and then with these heap it on to their hind-legs and knead it together. It is very worthy of note that in each flight they collect only one kind of pollen and bring it home unchanged, and by this means it is possible for the house-bees, as before related, to sort and carry it into different cells. This habit was remarked and described by Aristotle.

The nectar of flowers and all liquid foods are sucked up by the proboscis, which they push into the nectaries. In many of the flowers visited by them the nectaries are at the bottom of a tube which is partly covered and closed by the stamens. The bees find it out none the less, and if they cannot push the proboscis through the natural opening, they bite a hole, like the humble-bee, at the base of the corolla or even of the calyx, so as to be able to reach with their proboscis the place wherein nature has put the honey-gland. As the jaws of the bee are far weaker and more flexible than those of the humble bee, the former very gladly makes use, wherever it is possible, of the holes made by the latter in order to reach the honey, and only falls back on self-help when the other fails. On the other hand *Andrenetæ* and mason-bees, which also haunt such flowers, never utilise the holes made by humble and honey-bees, but try to get into the tube of the corolla from within—a fact which shows that these insects lack the sharpness of the honey-bee.

The collection of nectar is performed with great speed and cleverness, for we see the busy creatures fly swiftly from flower to flower. Arrived at home, to the interior of the hive, the flight or foraging bee deposits her outer and inner loads as quickly as possible, so as to return to the task of collecting, when, as before said, the provisions thus brought are properly divided and further worked up by the home-bees. If the foraging bee meets a hungry comrade on her way home, the hungry bee is sometimes seen, as with the ants, to communicate its want to its friend by touching its head with its antennæ, whereupon the latter opens its mouth and by regurgitating or disgorging the contents of the crop gives it food. Also when anything has happened at the hive so that the gathered provision of honey cannot be got at, or when such rich booty has been found that the superfluity cannot be taken in, similar scenes of mutual division may be observed, just as though, in order to provide against all the accidents of the future, the food at their disposal was equally divided among all.

In some wonderful way, in spite of the great number of workers, there is never the least press or the least disorder in going into or out of the hive. This may be explained either by each troop having special leaders which keep order, or by the bees which guard the entrance of the hive and have to warn off any unauthorised intruder, preserving order among the inflying and outflying bands. The bees maintain an efficient guard day and night during the warm season of the year at the gate of their dwelling, and it has to discharge many weighty functions. No one can penetrate into the hive without being first carefully touched and examined by this guard. As a rule, only inhabitants of the hive are let in, and all stranger bees, which are easily distinguished by the smell, are sent back. If a strange queen appears, the guard seize her on the spot. They grasp her feet and wings with their mandibles, and shut her up in so narrow a circle that she is unable to move and there can be no more talk about going into the hive. It only seldom happens that a queen, who has lost her way coming back from her wedding flight, penetrates by a badly placed or carelessly guarded mouth, into a strange hive, where she meets with inevitable death by hunger, smothering, or poison. The so-called robber bees, of which we shall have to speak again later, also sometimes succeed in mastering the guard and intruding into the hive by craft, strength, or fraud. But, as a rule, the latter are very much on the alert against these thieves and highwaymen, and very exceptionally let strange bees pass, and then only when they are laden with honey and pollen, and the guard are, therefore, convinced that they are not going to steal. In other cases, and when they have been warned and excited by previous burglaries, they rise several feet to meet the robber bees in the air, and try to kill them. They may then often be seen falling to the ground wrestling with each other. Only very young bees from strange hives, which have flown out and cannot again find the way to their own houses, are sometimes admitted, apparently from compassion, even though they are generally unladen. Even members of the hive during the foraging season are, as a rule, admitted only when they are laden, whereas the scouts returning from their expeditions are let in, as a matter of course, without carrying anything.

This foreign policy, as it is called, is not so strictly carried out by all bees as the bee-fathers might desire. Here, as in human policy, there is a good deal more connivance than is quite to be wished for in the interest of the community. But whether personal views, family and business connexions, hopes of advancement, fear of great folk, favoritism, etc., play a similar part to that they play among "reason"-endowed creatures, I do not venture to decide, though I think it very improbable.

The guards are chiefly on the watch against actual enemies of bees, or strange insects which try to penetrate into the nest. This happens most often by night, when the quiet prevalent in the hive favors approach, and the sweet scent proceeding therefrom attracts the foe. The instant that any strange insect comes into contact with the antennæ of the night-guard, the latter fly up, and instead of the short, broken hum that is heard when all is at peace, emit a very different sharp, hissing note, which is repeated by all the guards, and at once rouses the inmates of the hive. A number of working bees rush out and help to attack the foe.

Against such enemies as the guards are not able to drive away—such as the large death's-head hawk moth *(Sphinx Atropos)*, which is a great lover of honey, and in many years and places (such as Hungary) much infests the beehives—the clever insects manage to protect themselves by fastening up, or rather narrowing, the entrance of the hive with the propolis or propolis-wax, so that only the small bees can pass in and out, while passage is impossible to larger creatures. But they also defend themselves against smaller insects in the same way, because a narrow opening is easier to defend than a large. " I possess," says Jesse (" Gleanings in Natural History," vol. i., p. 21), " a regular fortress wall, built out of propolis, which one of my hives had placed in front of its mouth, in order to defend itself better against the wasps. With the help of this wall a smaller number of bees were able to defend the entrance."

If, however, a time comes when this narrow opening is no longer sufficient for the bees themselves, as, for instance, during a successful foraging, or when the hive is very populous, the formerly useful protection is torn down. Huber observed that a wall built in 1804 against the death's-head hawk moth was destroyed in 1805. In the latter year there were no death's-head moths, nor were any seen during the following. But in the autumn of 1807 a large number again appeared, and the bees at once protected themselves against their enemies. The bulwark was destroyed again in 1808.

It has already been said that animals, such as mice, slugs, etc., that have penetrated into the hive, are slain and then covered with propolis. Such a proceeding, however, was impossible with respect to a snail which, as Réaumur relates (Kirby and Spence, " Entomology," vol. ii., p. 229), took a journey over the sides of a hive lined with glass, its hard shell protecting it against the stings of the bees. The latter were equal to the occasion. They smeared round the edges of the shell with wax and resin, and fastened down the intruder to the wall of the hive in such fashion that he died from hunger or want of air! Sometimes animals which are so large that burying them in propolis is attended with difficulties, or that they would poison the hut in spite of it—

as, for example, mice—are gnawed to the bones, so that only their neatly prepared skeletons are found in the hive. The gnawed-off flesh, however, is not, as some have thought, eaten by the bees, but is carried out of the hive.

The guards have the further official duty of forwarding into the hive all news from without, and, according to de Fravière, they have a number of different notes in their voices, arising from the stigmata of the thorax and abdomen. As soon as a bee arrives with important news it is at once surrounded, emits two or three shrill notes, and taps a comrade with its long, flexible, and very sensitive feelers, or antennæ, which possess no less than twelve or thirteen joints. The friend passes on the news in similar fashion, and the intelligence soon traverses the whole hive. If it be of an agreeable kind—if, for instance, it concerns the discovery of a store of sugar, or of honey, or of a flowering meadow—all remains orderly. But, on the other hand, great excitement arises if the news presages some threatening danger, or if strange animals are menacing invasion of the hive. It seems that such intelligence is conveyed first to the queen, as to the most important person in the State.

This leads us naturally to the language, or the means of communication among bees, which language, although incomprehensible to us, is clearly capable of ready and distinct expressions. It is both a spoken and a gesture language, and there can be no doubt that by its aid the bees understand not only generalities, but also very distinct and very different things. The discovery of a treasure of sugar or other food in a pleasant place by a single bee has at once the result that in a short time a whole troop of hungry bees arrive thereat, and this can manifestly be owing only to a communication made by the first bee to its companions. According to Landois (*loc. cit.*, p. 153), if a saucer of honey is placed before a beehive, a few bees first come out, which emit a cry of *tüt, tüt, tüt*. This note is rather shrill, and resembles the cry of an attacked bee. Hereupon a large number of bees come out of the hive to collect the offered honey. When, in spring, the bee-master wants to make his bees notice the water placed near the hive—they need the water when the care of the young begins, for preparing the bee-bread—so that they may not be compelled to bring it from a distance, he need only hold a little stick smeared with honey at the mouth of the hive, and carry to the water the first few bees which settle on it. These few are enough to bring the neighborhood of the water and its exact place to the knowledge of the whole colony on their return. Herr L. Brofft relates, in the "Zoological Gardens" (XVIII. Year, No. 1, p. 67), that a poor and a rich hive stood next each other on his father's bee-stand, and the latter suddenly lost its queen. Before the owner had come to a decision thereupon the bees of the two hives came to a mutual understanding as to the condition of their two States. The dwellers in the queenless hive, with their stores of provisions, went over into the less populous or poorer hive, after they had assured themselves, by many influential deputations, as to the state of the interior of the poor hive, and, as appeared, especially as to the presence of an egg-laying queen!

The best means of mutual communication among bees, as among ants, lies certainly in their feelers or antennæ, with which they touch each other in many various ways. As the feelers are indispensable to them in all their work for arrangement and testing, no greater injury can be done them than cutting these off. The working bees subjected to such an operation became incapable of all work, and generally leave the hive in which they can no longer be of use. The drones also can no longer find their way about the passages to seek their food. They, therefore, also leave the hive in the darkness of which they have no guide. The queens lose with their feelers not only the consciousness of their maternal duties and the capability of discharging them, but also their feelings of mutual hate and jealousy. Antennæ-less queens pass closely by each other without recognition, and even the working bees seem to share their indifference, as though they were only apprised of danger menacing their nation by the excitement of their queen.

The best way to observe the power of communication possessed by bees by means of their interchange of touches, is to take away the queen from a hive. In a little time, about an hour afterwards, the sad event will be noticed by a small part of the community, and these will stop working and run hastily about over the comb. But this only concerns part of the hive, and the side of a single comb. The excited bees, however, soon leave the little circle in which they at first revolved, and when they meet their comrades they cross their antennæ and lightly touch the others with them. The bees which have received some impression from this touch, now become uneasy in their turn, and convey their uneasiness and distress in the same way to the other parts of the dwelling. The disorder increases rapidly, spreads to the other side of the comb, and at last to all the people. Then arises the general confusion before described.

Huber tested this communication by the antennæ by a striking experiment. He divided a hive into two quite separate parts by a partition wall, whereupon great excitement arose in the division in which there was no queen, and this was only quieted when some workers began to build royal cells. He then divided a hive in similar fashion by a trellis, through which the bees could pass their feelers. In this case all remained quiet, and no attempt was made to build royal cells: the queen could also be clearly seen crossing her antennæ with the workers on the other side of the trellis.

The bees, of course, make the greatest use of their feelers in the darkness of the hive or at night, while during the day and in the light they are guided by their rather short sight. In order to be convinced of this it is only necessary to follow their movements when they are on guard at the mouth of their hive on a moonlight night, to prevent the intrusion of the dangerous wax-moths fluttering above. It is very interesting to observe with what cunning the moths know how to utilise to the disadvantage of the bees the fact that they can only see things by a clear light, and what tactics the latter employ to discover and drive away their destructive enemies. Like watchful sentinels, the bees parade round their abode with constantly extended antennæ, moving them to right and left, and woe to the moth which comes in contact with them! But the moth tries to slip past them by carefully using every effort to avoid touching these sensitive organs.

Apparently the feelers are also connected with the exceedingly fine scent of the bees, which enables them, wonderful as it may seem, to distinguish friend and foe, and to recognise the members of their own hive, among the thousands and thousands of bees swarming around and to drive back from the entrance stranger or robber-bees.* The bee-masters, therefore, when they want two separate colonies or the members of them to unite in one hive, sprinkle water over the bees, or stupefy them with some fumigating substance, so as to make them to a certain extent insensible to smell, in order to attain their object. It is always possible to unite colonies by making the bees smell of some strong-smelling stuff such as musk.

The remarkable memory of the bee is undoubtedly connected with the acuteness of this sense, which enables it to find again the old foraging places, the tree or the flowers where it has once found honey, and which make it possible for it to recognise its own hive among many others. Huber relates that one autumn he put some honey in a window to which bees came in crowds. The honey was taken away and the shutters remained shut all through the winter. When they were reopened in the following spring the bees also returned, although there was no honey in the window. They, therefore, without doubt, remembered that which had before stood there, and the lapse of many months had not erased the impression.

Stickney also relates a remarkable instance of bee-memory (Kirby and Spence, *loc. cit.*, Vol. ii., p. 591): "Some bees which had taken possession of a hole under a roof, but were removed into a hive, sent out scouts from their new home to this hole for several years in the swarming season. The remembrance thereof must therefore have been transmitted from generation to generation, or communicated." In similar fashion Karl Vogt ("Lectures on Useful and Harmful Animals"), states that ants for year after year went through several inhabited streets to the store-room of a chemist 600 metres off, in which stood a large vessel always filled with syrup.

---

* According to the late investigations of Dr. O. J. B. Wolff, in Coswig, near Dresden, the smell of the bees does not lie in their feelers, but in two special smell organs, situated near the pharynx, consisting of 110 pairs of papillæ to each of which goes a special nerve. The insects allied to the bees, wasps especially, have only from 20—40 pairs of these papillæ. So there we have the explanation of the wonderfully fine smell of bees, and of many of their faculties, considered inexplicable till now. Between the large compound eye and the root of the upper jaw lies the mucilaginous gland of smell, which secretes the mucous which increases the sensitiveness of the organ and keeps the smell-papillæ moist.

The certainty with which bees find their way back to their hive after foraging proves their excellent memory. As a bullet from a gun they fly by the shortest way to their beloved home, for instance, at the sudden approach of a storm. This power of finding their way back has indeed a limit, and it is said that bees which have gone to a greater distance than half-an-hour or an hour from the hive, easily lose their way on returning. Therefore they are the fonder of a flowering field or similar spot the nearer it is to the hive, even omitting that such nearness spares both time and strength. Perhaps they fear sudden gusts of wind so much, as was before mentioned, because these drive them far enough away from home to make return difficult or impossible. Whether, as Virgil says in his famous poem on bees, in places where the wind threatens them dangerously, they try to save themselves by lifting little stones and bits of earth from the ground with their feet—and in this way oppose a better resistance to the waves of the air, just as a ship laden with ballast resists the waves of the sea better than an empty one—is not yet certain. But let Virgil himself tell his observation in connexion with his description of the bees' flight:—

> "At morn like soldiers pour they from their gates,
> And not a bee behind then idle waits,
> But when the eventide now warns to stay
> Their gathering honey on the plain, then they
> Fly homewards, and fresh strength from food derive.
> A buzzing rises, and around the hive
> And by its entrance-door they hum; but when
> They've settled on their couches, all is then
> Hushed for the night's reposo, and kindly sleep
> Comes o'er their weary limbs. They ever keep
> Close to the hive when rain-clouds low'r on high;
> When east winds blow ne'er do they trust the sky,
> But safe beneath their tiny city's towers,
> And round their home fresh water from the showers
> They get, or short excursions try, and buoyed
> Oftimes by little pebbles, through the void
> Of heaven fly steadily; so rocking boats
> In tossing seas take ballast in."
> [*Loc. cit.*, pp. 28, 29.]

No less poetically, but more briefly and more strikingly than Virgil, the great poet Shakspere describes the well-ordered life and doings of the bee State, putting the following into the mouth of the Archbishop of Canterbury (from the standpoint of the absolute Prince of the Church) in his drama, "King Henry V.": [Act 1, Sc. 2.]

> "Therefore doth heaven divide
> The state of man in divers functions
> Setting endeavor in continual motion;
> To which is fixed, as an aim or butt,
> Obedience; for so work the honey-bees;
> Creatures that, by a rule in nature, teach
> The act of order to a peopled kingdom.
> They have a king and officers of sorts;
> Where some, like magistrates, correct at home;
> Others, like merchants, venture trade abroad;
> Others, like soldiers, armed in their stings,
> Make boot upon the summer's velvet buds;
> Which pillage they with merry march bring home
> To the tent-royal of their emperor:
> Who, busied in his majesty, surveys
> The singing masons building roofs of gold;
> The civil citizens kneading up the honey;
> The poor mechanic porters crowding in
> Their heavy burdens at his narrow gate;
> The sad-eyed justice, with his surly hum,
> Delivering o'er to executors pale
> The lazy yawning drone. I this infer—
> That many things, having full reference
> To one consent, may work continuously;
> As many arrows, loosed several ways,
> Come to one mark."*

---

* In Virgil's and in Shakspere's times the queen was regarded as king.

CHAPTER XXI.

MONARCHY, SOCIALISM, AND INSTINCT.

THE bee State has often been held up as the ideal and example of the system of so-called constitutional monarchy, of that system which is now prevalent in most European countries, and which is regarded by some as the highest political ideal, and by others as a gross political sham. The Frenchman, Mandeville, as long ago as the beginning of the last century, in his famous (or much talked-of) "Bee Fables," held up the polity of the bees as a model for human arrangements, although in a very exaggerated way.

As a matter of fact there is no small resemblance between the bee system and that of constitutional monarchy in so far as the bees appear to lay no stress on the person of their queen, and are perfectly contented so long as they have one, that is someone capable of discharging the royal or rather maternal duties. They change the sovereignty, as a rule, easily and quickly, and thoroughly admit the well-known maxim of constitutional royalty: "*Le roi est mort—vive le roi!*" (The king is dead—long live the king!) A chiefless hive, one robbed of its queen, either does homage, as already described, to a fresh queen introduced into it just as to her predecessor, or brings up a new sovereign by its own efforts; while a hive, left long queenless, falls into sloth and riot and sooner or later perishes. The queen, since all revolves round her, is the necessary centre and bond of the hive, but without herself taking any personal part in the business and proceedings. She therefore, in reality, exactly answers to the foundation-stone of constitutionalism, and is what Napoleon I. declared he would not be, in reply to the famous constitutional reproach of Sièyes: "The prize pig of the nation." She is indeed widely separated from her human antitype in that she is not merely "representative," giving to high and low merely an empty show, but really discharges actual and essential duties, without which nothing could exist. Apart from this, the queen in the simplicity and uniformity of her work, and in the half, though respectful, imprisonment in which she is kept, is a complete contrast to her intellectually and physically developed and active subjects, so that here, as so often among men, it appears fair to say that stupidity or narrowness, or perhaps only mediocrity, rules over reason.*

In any case this sovereignty, as we have seen, is much restricted by the subjects themselves, and these seem to indemnify themselves for the compulsory endurance of a monarchical head, by observing amongst themselves, on the other hand, the maxims of the most extreme democracy, of the widest Socialism and Communism. One is as good as another, the beautiful principle is unconditionally obeyed: "One for all—all for one." They have no private property, no family, no private dwelling, but hang in thick clumps within the common room in the narrow space between the combs, taking turns for brief nightly repose. The building, cleansing and working are also carried on partially all through the night. All stores are common; there is only the state magazine, and all are fed from this without distinction of person. If want and hunger enter, all die alike. The queen here is an exception and has the privilege of dying last. The bees are, however, egoists enough in such times of need, or in threatening famine from continued bad weather, to throw the larvæ, the drone larvæ first, out of the cells. This also happens, on the other hand, when lack of place for storing provisions occurs, owing to very successful foraging. The larvæ are then thrown out, or the nursing narrowed down to the uttermost.

---

* Espinas ("Animal Communities") protests against the expressions "monarchy" and "queen" as descriptions of the bee State (although these expressions have always only been used figuratively), because the queen exercises no real sovereignty, but only acts as mother, and the workers are not subjects, but help as foster-mothers and nurses. He forgets that in the true constitutional system the human being also does not rule, but only acts as the central point of the State, which lends him harmonious support, and thus has the greatest resemblance to the position of the queen in the bee State. Espinas himself admits this in another place, by calling a beehive a "moral organism," or a real consciousness, whose leading idea, or chief part, is the mother.

In matter of labor the bees have realised the highest ideal of Communism, for it is perfectly free, voluntary and uncompulsory; each does as much or as little as seems to it good. But there are no sluggards among them, for the universal example acts as an incitement, and in a society wherein all work idleness is really an unthinkable and impossible thing, whereas on the contrary, in the much praised opposite condition of human society, the idleness of the few is not only favored, but seems to be absolutely unavoidable. Truly, in a communistic form of society the individual must have the consciousness, as among the bees, that in so far as he is a member of the whole, he is not working for others, but for the common good and therewith for himself. This consciousness makes the bees such busy and eager workers, that many of them work themselves to death in a few weeks during the foraging season, whereas working bees usually reach an age of nine or ten months, so that Virgil wrote truly:

"Ofttimes in a mistaken flight they tear
Their wings, and even generously die
Before they drop the precious load, so high
The fame of getting honey, and so strong
The love they feel for flowers."
[*Loc. cit.* pp. 29, 30.]

The "instinct"-philosophers will probably say that this is only the result of an inborn irresistible heaven-implanted tendency in the little bee minds, from which the insect cannot voluntarily free itself, and that we therefore cannot here speak either of merit or design. But in the first place it is not credible that instinct should dictate to an animal to do that which will finally lead it to destruction, and secondly that opinion agrees very ill with the already often-mentioned experience, that the inhabitants of a queenless hive, which with their queen have lost the object of their society, cease to work and fall into idleness and riot. These, in consequence of an event quite outside themselves, the connection of which with their personal life-duties can only be clear to them by a decided act of reflexion and deduction, have entirely lost the impulse to work, formerly so strong in them, and this is impossible and incredible if this tendency or impulse were instinctive, or unconnected with the volition of the individual. The members of queenless hives scatter and die, or try to slip into other hives to get food, wherein they generally do not succeed, being sent back by the guard. The Australian experience also—that bees taken there from Europe quite gave up storing honey for the winter after the lapse of a few years, because they learned by experience that the continual summer rendered such action unnecessary, while they otherwise kept their hives in the most perfect order—is irreconcilable with the idea of an inborn or irresistible work instinct.

This opinion comes to sad shipwreck on the so-called robber bees, which try to lighten or quite to spare their labor by falling in crowds upon other already-filled hives, mastering the guards and the inhabitants, robbing the hives and carrying off the stores thereof to their own dwelling. If this plan has succeeded once or more, they, like men, find more pleasure in robbery and in plunder than in their own work, and become at last formidable robber states. Single bees also go out robbing, and try to penetrate into a strange hive undiscovered and in cautious fashion, showing by their whole behavior that they are perfectly conscious of their bad conduct, whereas the workers belonging to the hive fly in quickly and openly in full consciousness of their right. If the solitary robbers and sweet-stealers succeed in their attempt they lead other bees from their hive into the same theft. These are always followed by more tempted ones, so that finally the robber state is formed. The bee-masters, therefore, in order to avoid injury from robbery, are obliged to put an end to the mischief as soon as possible, before the bad example has infected others. The inhabitants of an attacked hive naturally defend themselves with all their strength, so that the burglary generally is only successful in sick or weak hives. In powerful, well-organised communities, the robbers are, as a rule, flung back by the guards and pursued. If, however, they find a hive from the mouth of which they are not driven away, but are let to slip in, they eat the honey, carry it off to their hive, manifest their delight at their own entrance, and offer their proboscides to their sisters to let them taste the new store. They soon return in increased numbers and more eager than before, and try to get into the hive in every way, by crevices and so on. Arrived within, they try before all else to kill the queen, for they well know that the assailed hive will then lose unity and power of resistance, and that the inhabitants will weakly surrender on the death of their leader. Bees from other hives also join the invaders, and the end is thorough robbery and plunder, which is all the more complete as the owners of the plundered hive when they find that all is lost and that no further resistance is possible, generally themselves join the thieves, tear down the cells, plunder them, and then go off to the robber-hive.* When the assailed hive is emptied the next ones are attacked, and if no effective resistance is made are robbed in similar fashion, so that in this way a whole bee-stand may be gradually destroyed. Sometimes the resistance of healthy hives fails owing to the robbers, from visiting the same flowers or the same fields, having the same smell as the bees of the attacked hive, and therefore not being recognised at once as thieves.

They sometimes then become so bold that they post themselves in front of the hive, often stop the bees returning from collecting, which, as a rule, or very often, rest for a brief siesta on the stand ere entering the hive, and partly by threats, partly by force, deprive them of their load of sweets. E. Weygandt, who personally observed this interesting kind of robbery, and described it in the journal *The Bee* (1877, No. 1.), called it "milking," and states that this milking has been observed by many other bee-masters. The milking bee, according to him, gains the further advantage that with the honey of the milked one it has also contracted its smell, and partly owing to this, and partly to the fact that it arrives loaded, it is admitted without difficulty into the hive, and can prosecute its thefts. The robber bees in this resemble swindlers who dress themselves like policemen, and carry out their schemes under this mask. Sometimes the bee-masters defend themselves against robbery by putting musk in a plundered hive. The robbers then contract the musk-odor, and when they go back to their hive they are regarded by their own comrades as strangers, in consequence

---

* Siebold saw the same kind of thing with the French wasp (*Polistes gallica*), of which mention will be made further on. Stranger wasps attacked one of its nests, tore the larvæ out of their cells and carried them home as booty. When the real owners of the nest saw that resistance was vain, they followed the example of the robbers and murdered their own children. (In Graber, *loc. cit.* ii., p. 134.) "Which is this," the reporter adds to the story, "'instinct' or 'unconscious intelligence'?"

of the smell, and are driven away or killed. Most of the robberies take place out of and soon after the collecting season, because the bees flying about and accustomed to forage can find no more food nor provisions, and therefore seek other opportunities of gain, be they fair or foul.

In addition to these robberies from their own race, there are a large number of somewhat similar opportunities of theft, and the bees thoroughly understand how to utilise these cases with great craftiness. Instinct is here quite blameless, for most of these incidents are quite accidental, and brought about first by man's industry. The sugar plantations in Cuba (and in other places, as in the neighborhood of Stettin, where there are many sugar refineries, and where it is asserted that the bees know very well how to distinguish the different kinds of sugar) suffer no little injury on account of the visits paid by the bees to the refineries. Where such resources are open to them all the year through, as in Barbadoes, they finally lose their original instinct of work, and disuse the collecting of honey. They also do not hesitate to take the wheat or rye meal placed as food by the beemaster in front of the hives in spring, before the flowers begin to bloom, and turn it to account as a substitute for the missing pollen.

Amongst us also, during the late summer and autumn, when the flowery food supply begins to run short, the bees utilise every opportunity of stealing sweet things, and it is well known to everyone that confectioneries, sugar factories, and similar places are regularly besieged by them. They trace out every such spot with tireless patience, however concealed it may be, or however difficult to get at, as, for instance, syrup-casks in cellars, which can only be reached by narrow cracks in the cellar shutters. Bee-masters thereby suffer great loss, for on such occasions many bees perish for the same thing through which so many men lose life or health—namely, through intemperance. They drink so much that they fall to the ground, and are no longer able to return home.

Nor do they neglect such chance opportunities as may be offered them by nature. They are as fond of the honey collected by humble bees as of their own, and have many sly ways of getting hold of it. Huber, during a season of scarcity, put the nest of some humble bees in a box near his own bees, which promptly ravished it. Some humble bees which, in spite of the disaster hovering over their nest, had remained therein, continued to fly out and to bring food for their wants back to the old place. The bees followed in their tracks, turned back with them into their nest, and did not leave them until they had obtained the surrender of their harvest. They licked them, stretched out their proboscides to them, surrounded them, and would not let them go until they had won from them the sweet juice which they concealed within them. They did not try to kill the insects to which they owed their meal, while the good-natured and rather stupid humble bees, which became quite accustomed to the payment of these contributions, gave up their honey and flew out again. This new style of housekeeping was carried on for more than three weeks, until at last the humble bees scattered, and the beggar bees consequently came no more. Some wasps, which paid a visit resembling those of the bees, did not succeed. They could not make friends with the original owners of the nest in the same way, and did not possess the fine coaxing or sly manners of their rivals.

An exactly similar scene may be observed between robber bees and the inhabitants of a weak hive, and strikingly recalls the details given above.

Robber bees can be raised artificially by feeding bees with honey mixed with brandy. They learn to be sadly fond of this drink, just like men, like them, soon become stupid and intoxicated, and, like them, cease to work. If they then become hungry, they fall, like men, from one vice to another, and take to thieving. Instinct withholds them from this fatal enjoyment just as little as it does from the enjoyment of bad or sour honey, by which last proceeding bee-masters often suffer great loss. According to the newspapers, no less than 550 hives perished in Boone County, in America, during April and May, 1872, in consequence of the bees eating sour honey.

All this, and many other similar facts, show that the bees in their actions do not, as has usually been thought without proof, obey a distinct irresistible impulse of nature, but that with them, exactly as with men, work and pleasure are distinct and various, according to the difference of circumstances and of wants. " How shall we name," says A. Fée (*loc. cit.*, p. 108), this care and thought for each individual case and condition? this remarkable division of labor? this wonderful polity, which orders everything according to settled rules, and momentarily meets a mass of accidental circumstances which it was impossible to foresee? The bees feel anxiety, hate, and wrath. They accommodate their actions to circumstances, they employ stratagems against enemies which are stronger than themselves, and suit the defence to the strength of the attack. Can this be instinct?"

"To deny reason to bees," says Leuret, "is. to deny all justice!"

CHAPTER XXII.

Cell-building.

THIS same point of view must be taken, as for all else, in studying the famous pyramidal and hexagonal cell-building of the bees already described, although in this people have mistakenly tried to find an irrefutable truth of an intelligence and a mathematical knowledge impossible in them. We have seen that the bees often build cells of other shapes as well, and that they mutually correct and improve each other in the building itself. We know that they utilise, for the support and upholding of their combs, bits of stick artificially introduced into the hive: yes, that they will use artificially-prepared combs, or old ones emptied of their honey by human hands, as though they were their own: or that they will continue combs artificially begun, as though they themselves had commenced them.* Different races of bees, such as German and Italian, show considerable difference in their work, but it is surely impossible that there can be a German and an Italian instinct.

But with all this, the remarkable shape of the individual cells, so exactly agreeing with geometrical laws, is not itself explained, and on this point we should be obliged either to confess our ignorance or to allow that the believers in instinct are right, if we were compelled to admit that the bees have always, since the first origin of their race, built such cells as we see to-day. But the great and important law of gradual evolution, to which the whole organised world owes its origin, here again solves the problem for us, and teaches us that the present shape of the bee-cells has arisen

---

* If a plate of wax, with artificially-made commencements of cells of the size and shape of ordinary bee cells, be hung in a hive, the bees will build on this artificial foundation; bee-masters utilise this to shorten the long introductory labor in the hive, since such manufactured articles have been sold.

gradually in perfectly mechanical fashion from pressure of space and mutual flattening from the originally imperfectly shaped cells, and that the need of economy in space and in wax was the impelling reason for the prosperity and propagation of such hives as gradually advanced to the conception of the perfect cell form. That this is no theory, but fact, is proved by the transition shapes and gradations between perfect and imperfect cells among the nearest relations of the honey bees, such as humble bees, mason bees, *Anthophora*, *Melipona*, wasps, etc., which are met to-day in great numbers and variety. According to Darwin, the humble bees are at the extreme end of the scale on the imperfect side; these use their old cocoons or pupa-cases for the reception of honey, sometimes adding short wax cylinders, and, in addition, making a few separate irregular rounded cells of wax. These cells, which may best be compared to eggs with the point cut smoothly off, or to the open end of a narrow thimble, generally lie irregularly together, or are at the best laid on a short horizontal platform, raised on pillars, so that a humble bee's nest in comparison with a beehive is, to borrow Réaumur's expression, like an irregularly-built village compared with a well laid-out city. "Regular order, beauty, and grace of form are just as little to be found in our towns, where this house stands one way and its neighbor the other" (Giebel).

Between these imperfect nests and cells of the humble bee and the perfect ones of the hive, or honey bee, there is a countless number of gradations among the various species of bees, as well as among their nearest relations the wasps, with their numberless species and sub-species. Darwin notes among these gradations as specially remarkable the architecture of the Mexican *Melipona domestica*, an American species of bee, which makes an almost regular waxen comb in the cylindrical cells wherein the young are tended. It also builds for the storage of honey a number of larger cells of almost spherical shape, and of nearly the same size, which are pressed so closely together that at the places of contact the rounded form is lost, and a flat sheet of wax forms the partition-wall instead; this is the beginning of a mutual flattening of the previously spherical cells. If the *Melipona*, like our hive bee, made its spherical cells of the same size and at given equal distances from each other and symmetrically in a double layer, and tried to save as much space and wax as possible, its structure would then be almost as perfect as that of the hive bee. Let it now be remembered that the bees always begin with circular holes in a thick wall of wax, and later erect the single partition-walls,- and in order to make the greatest possible room within for the reception of honey and to use as little as possible of the costly and precious wax, they work out the corners sharply, so that each bee carries on its gnawing to the highest admissible extent of thinness of the partition-wall; let it further be remembered that the mathematical exactitude of the cells has been much exaggerated, and that some are very regular and others very irregular, and that some five and four-cornered cells are found, and we shall then easily come to the conclusion that early ancestors of our honey bees built in as imperfect fashion as the *Melipona* and only gradually worked on to a perfect system of architecture.* In any case we must disagree with those who think that there is nothing to show such an improvement in our honey bees, and that these rather build their cells to-day exactly as they built them two or three thousand years ago, and as they will probably build to all eternity. But—apart from the fact that it would be exceedingly difficult to prove the justice of this contention—it is forgotten that bees are not three thousand, but hundreds of thousands of years old, and that on this road they have long attained a grade of perfection which is thoroughly sufficient for their wants and therefore cannot be improved. What Häckel ("On Division of Labor," 1869) says in this connexion of ants is quite as true of bees: "These rude aboriginal ants, which lived many thousands of years ago, perhaps even during the chalk age, had as little idea of the advanced division of labor of the various modern ant States as had our German ancestors of the Stone Age of the high culture of the nineteenth century. These,

---

* According to Graber (*loc. cit.* ii., p. 78) Heinrich Müller has shewn it to be in the highest degree probable that our present bees are descended from certain digging wasps, from those robber *Hymenoptera* which themselves live on flower-juices, but feed their young, brought up in holes, with insects. Gradually they came to substitute vegetable matter for the flesh-food which was often only to be obtained with difficulty.

like those, have slowly and gradually struggled up the toilsome road of progressive development. Even now there are some species of ants ignorant of the highly developed division of labor in the civilised ant States, and which bear the same relation to these as the rude aborigines of Australia and Africa do to the civilised and cultured nations of to-day."

In this connexion the observations made by Bates (*loc. cit.*, Vol. ii., p. 44) on the American bees, or *Meliponæ*, are very noteworthy. It seems, he says, that none of the American bees have reached that high stage of architectural ability in the formation of their combs which is seen in the European honey bee. The wax cells of the *Melipona* are generally long and only show an approach to the hexagonal form where many of them come into contact. This certainly shows plainly enough how the purely mechanical force of pressure and narrow space must gradually cause the originally spherical cells to change into cornered, and especially six-cornered ones. For the hexagon is exactly that geometrical form in which it is the most natural for little bodies, not laid above each other in rows with mathematical accuracy, to unite without gaps and interspaces. Therefore such bodies, as a matter of fact, when they are soft, vesicular or yielding, flatten themselves into hexagons of their own accord in a given narrowed space. For instance, if enough water is poured into a bottle filled with peas to make the peas swell, and they are not able to get out of each others way, on emptying them out hexagonal and not spherical bodies will be found. We have the same appearance if we blow air into soap-suds. All the rising bubbles are pressed together in more or less six-sided forms, whereas bubbles set free in the air are perfectly round. Also the originally spherical cells of which our own body is composed assume the hexagonal form wherever they are closely pressed together, as in mucous membranes, cancerous tumors, etc. Now let us imagine two flat layers of equal-sized thimble-shaped cells, like those, for instance, built by the *Melipona scutellaris* (compare the picture of p. 464 in Blanchard's work), so placed one above the other that the openings of either layer point outwards, and that the closed ends touch each other, so that each convex surface of the one side fits into the concavity made by the ends of three contiguous cells of the other, and imagine that the whole somewhat yielding cells are subjected to a gentle mechanical pressure from above and from below, and we shall then necessarily obtain the shape within and without offered now by the appearance of a double-sided honeycomb; that is, the cells will each have bent into hexagons, and at their closed ends into small three-sided pyramids with the much-admired dove-tailing of their three rhomboid sides. The already mentioned gnawing away of the several cells in order to save the wax does the rest. It must not be forgotten also, that our honey bee, like the *Melipona* species named, sometimes builds one-sided combs, and that a double-sided comb is not thinkable in any way save just in the fashion in which the honey bee builds it.

Above all things it must not be imagined that this mechanical reason for the building of the cells of the honey bee is now necessarily practised. The reason is long gone by, but the work has remained, and this the more because the important matter of sparing space in the narrow artificial hive in which the honey bee is now compelled to live, almost without exception, appears of double significance. Each single bee now at once builds, without knowledge of the reasons which during the lapse of ages have brought about the particular shape and boundaries of its cell, in a way which is prescribed to it partly by its inborn and inherited tendency, partly by the size and shape of its body, partly by the conscious design of saving space and wax, partly finally by the instruction given by older comrades. As a rule people are little inclined to admit the last influence, owing specially to the extraordinary brevity of a bee's life, which does not generally last for more than a year, although Virgil makes it live during not less than seven summers.*
But it is forgotten that with this short-lived and most industrious creature a day of life is the same as a year to men, and that if such education takes place it would progress with extraordinary rapidity. We have also seen

---

* The queens live from three to four years. It is therefore quite possible that working-bees may sometimes reach a greater age, although most, according to the opinion of bee-masters, do not survive more than half a year, and many, as already said, work themselves to death in the course of weeks or months. The drones as a rule do not live more than a quarter.

T

among the ants that their whole education is completed in the course of a few days.*

There is indeed a difficulty touching the inborn and inherited building tendency of the bees, or possibly of an inherited idea of a certain cell-form (which last opinion will not be dealt with further owing to the great obscurity which still surrounds the facts and laws of psychical and psychical heredity), which might be absolutely fatal to such a supposition. How can we speak of inheritance, it is asked, among creatures which, like the neuter bees and ants, or like the merely working members of the insect colonies, close the cycle of their whole existence with their personal activity, without being able to bequeath their acquired capabilities, habits, or talents to their posterity? while, on the other hand, such a transmission is impossible on the part of the real parents, the non-working fertile females and males of the colony, which are far behind the workers in intelligence and skill!

A glance at the historical origin and the past of the bee nation will give the answer to this apparently difficult question. For there can be no doubt that, as already partly explained, the present organisation of this State and especially its highly developed division of labor, which saves the real founders of the colony from all work, is only the gradual and slowly-ripened product of historical evolution, and that the organisation has not always been that which it is to-day. We find transitional steps here in large numbers, as in the cell-building, among the nearest relations of the bees. The fertile females and males work among the solitary or unsocial bees, as well as among humble bees and wasps; indeed by far the greatest and most important share of the work falls to the lot of the females. The female wasp herself builds her nest and her cells in the spring, lays her eggs in them, and nurses and feeds her young until the later-emerging workers can relieve her of her heavy work. But even then the female is ceaselessly busy, while the males, emerging at the end of August, clean the nest and carry out excrements and dead bodies. The female humble bee is not less industrious, but works in spring with such rapidity and

---

* Further details on the gradual development of the cell-building instinct of bees will be found in Graber (*loc. cit.* ii., p. 178, etc.).

skill that she makes a larval cell, fills it with honey or bee-bread, and lays an egg therein in the course of half-an-hour. The later-emerging fertile females and males help the mother in building the cells and in tending the brood. The other solitary species of bees act in a similar way. The female of the mason bee (*Megachile muraria*) builds in the spring her thimble-shaped larval cells, just as do swallows, out of earth or sand kneaded with saliva, on the sunny sides of garden and stable walls, and this all by herself, although this kind of building requires much skill, industry, and patience. She lays an egg in each cell, after she has put therein a jelly of pollen and honey as food for the larva after hatching. The cell is then closed, another built in similar fashion, and so on. The partition-walls are also cemented, so that the individual cells may have more cohesion, and the whole is covered with a protecting roof of rather rough mortar. We have seen among the ants that although the queens do not work as a rule, they are yet very well able to do so, and that there is indeed a single species in which they regularly take part in the work. We have also seen them take a share of fighting, and that in very powerful and effective fashion. It was also shown that some fertile females after the wedding flight dug holes in the ground, and just like wasps and humble bees founded new States and colonies independently and without foreign aid, whilst the founding of new colonies was generally done by emigration from over-populated States. It seems as though the apparent intellectual sloth or inferiority of the fertile females of the bees and ants in comparison with their working sisters is *only* apparent, and is grounded on the difference of their work. At least the sensible conduct of the queen-bees in occasional emergencies of life, as already described, is in favor of such a conjecture, as is also the founding of new colonies by individual female ants, or their occasional participation in work or in battle.

In view of such facts there is not the smallest reason to prevent us from believing that the female bee originally formed her own colony and was at once queen and worker in her own person, just as are to-day her relations above-mentioned. The now so idle drones may also in long past times have rendered services which were later taken from them by the industrious workers as the division of labor progressed. But although the queens and drones do not now work yet the capacities inherited from earlier times still remain to them, especially to the former, and are kept alive and fresh by the impressions constantly made upon them during life, and they are thus in a position to transmit them to their posterity. In any case the now so firmly established condition of the bee State, resting as it does on a steady historical past, agrees with this idea of capacities acquired by the force of heredity, but no longer capable of further development or perfectibility.

The views here expressed become almost a certainty when we learn from Graber (*loc. cit.* ii., p. 88, etc.) that the queen, although owing to the division of labor in the bee State she has no longer need of it, still possesses the collecting baskets on the hind legs—so indispensable to the working-bees—as a rudiment or remainder from the earlier time when she was an ordinary worker, while on the other hand, as before-mentioned, the workers possess rudimentary ovaries. "The worker," adds Graber, "is thus a mutilated queen, and the queen an emancipated worker." Therefore again, as we have mentioned before, is it possible for workers occasionally to set themselves up as queens, doubtless in remembrance of their former more exalted position!

That this opinion, if correct, would also apply to the other social insects, and especially to ants—supposing that the force of heredity in explaining their capacities and habits cannot be denied—scarcely requires special argument. But that these capacities and habits, when once present, should be similarly repeated in each new colony cannot seem strange, owing to the planting of new colonies from the old or by the mothers. The young bees and ants have only in all cases to follow the example of the older ones which they have before them, and the faculty of imitation is, as we see by countless examples, a chief characteristic of common work in all social insects, as among men.

But to return once again to the constitution and arrangements of the bee State, it must be admitted, on unprejudiced observation, that the idea of a well-ordered State, politically and socially, is here indeed almost attained. There is among them no standing army, as among other insects related to them and among men, but the State seeks protection (in cultivated hives become partly unnecessary) against outside attacks in the universal arming of its working citizens, just as the burgesses of towns in the Middle Ages were at once craftsmen and soldiers. In the interior it is work only, and that of the most selfless kind, always directed to the common good, which is the bond of union. This work, and this absence of an army always prepared for war, prevent the bees from thinking unnecessarily about foreign wars and regularly organised pillaging expeditions and slave hunts, like the ants. Wars which somewhat resemble those of the ants only arise when the home is to be defended against foreign invasion and attack, and especially against robbery by members of their own race. A furious battle sometimes also breaks out when two swarms with different queens meet, possibly because each swarm fears that the other wants to steal its queen. As a matter of fact the battle as a rule ends when one of the queens of the hostile bees has been killed. It is not yet explained why, as Scheitlin relates, a kind of civil war and a general hand-to-hand fight sometimes arises in the interior of a hive, and ends with the death of many; possibly robbery may here also be the inducing cause. Private quarrels also appear to occur. These, however, must not be fought out in the hive itself, but outside, and as a rule end in the death of one of the duellists, which has received the stab of the terrible poisoned sting between the segments of the abdomen.

Whether, as Virgil so poetically relates, quarrels between single queens can cause wars and battles between whole hives and swarms is doubtful and not probable, when we remember that the workers, as a rule, leave the queens to fight out their own quarrels, and only take part as passive spectators. It is more likely that such battles should take place for the valuable possession of a queen. Yet it is no uncommon sight for two swarms in which there is only one queen, instead of fighting, considering it better for the common interest to utilise their forces by uniting together. Perhaps the bees, since Virgil's time, have become more peaceable and more sensible on this head, and have recognised, better than do men, that war is the greatest evil and the greatest folly on earth, especially when waged in the interests of the ruler and not in those of the people.

Our admirable bee democrats must not be too severely blamed for their " crowned head," when we remember how thoroughly overlooked the queen is by the workers and how dependent she is upon them, and that her sphere of authority does not nearly approach that of the president of a human Republic. She appears less as ruler than as the first and most important servant of the commonwealth, and homage is paid her more from love and respect and because the existence of the colony depends upon her, than from fear of her sovereign power. Nor has she, like the constitutional human monarch, the remarkable prerogative of personal immunity and irresponsibility, but must give throne and life as security for the fulfilment of her royal duties.

Statesmen, workmen, and reformers, take ye an example therefrom!

But not only in political but also in personal training may the bees serve us as models and examples. For where shall we find united with so much virtue, industry, and self-sacrifice, so much modesty and simplicity of form and of behavior? What a distance between the quiet bee in its simple dress, and the glittering thousand-hued butterfly, a coquettish idler, fluttering from flower to flower, from pleasure to pleasure, and delighting the eye of the spectator with its wealth of color! or the proud buzzing and humming beetle, letting the rays of the sunlight glance back from the golden and glittering wing-cases! And although they both draw the eyes of the world on them and are admired and sought by all, yet how far they stand below our bees in intelligence and accomplishments, for which they only can feel admiration who know them, and who are able to judge their merit and to prize them. What a wonderful picture of human life and common-place human estimation! The thoughtful Greeks indeed, with a warm fondness for the famous honey of Hymettus, have shown excellent views of true virtue and true merit by calling their God the Father, Zeus, the God of Bees and the Father of Bees, and by making sacred bees guard the grotto in which he was born. In the form of bees the Muses showed to the Ionians the path from Attica over the sea to Asia, and priestesses were named bees on account of their peculiar sanctity. Born of the sun, the bee ever remembers its home, while the lazy drone is born from the carcase of the horse. Bee-souls are therefore those souls which keep themselves pure and think of results. Bees avoid all that is base. They placed sweet honey on the lips of the new born Zeus, and rejoiced over him, and the gods on high Olympus feast on honey in nectar and ambrosia (Scheitlin, *loc. cit.*, p. 115). Perhaps also it was the bees which suggested to the Greeks and their great poet Hesiod the deep proverb, that the gods had placed toil before talent! They certainly might have done so. And when Pliny in his "Natural History" (Book xi.) told of two Greek sages (Aristomachus of Soles and Philiscus of Thasus) who had devoted all their lives to the study of bees, this further proves how rightly the Greeks knew how to value the interest and merit of this wonderful insect. This interest has not only endured down to the present time, but has increased with the more intimate knowledge of the domestic life of this remarkable creature. "From bees," says Dr. Dzierzon, of Carlsmarkt (*loc. cit.*), to whom we owe so much of our nearer knowledge of this life, "since their activity, their whole domestic life has latterly been completely revealed to men, man can learn even more than from the ants, set out in the Scriptures as an example shaming the sluggard. The industry of the bee is unwearying and it often falls a sacrifice thereby to cold air. In its cleanliness, mutual attachment and compatibility, its disinterestedness, dividing the last drop of honey with its sisters, in its tender love for the common mother and ruler, in its courage in defending her and its hive, rushing against an enemy threatening its destruction with a real contempt of death, the bee is to man a teacher of the fairest domestic and civil duties. If each citizen of a state acted from conviction and a feeling of duty as the bee acts from direct inspiration or instinct, such a state might well call itself happy."

Our forefathers also, the ancient farmers, valued bees highly, on account of the mead which they prepared from their honey, and unusually long and broad honeycombs were brought out of Germany to Rome. They did not keep the bees in wood or straw hives, but in hollow trees, such as the wild wood bee, of which we shall speak presently, still uses for its nests.

CHAPTER XXIII.

Other Species of Bees.

THE last-named bee, like all the other species of bees or *Apides*, of which there are several hundreds, are far behind the European hive or honey bee in intelligence, although much that is wonderful and surprising is related of many among them, and although both their organisation and their habits closely resemble those of the honey bees. They also, without exception, show great skill in building. The *Osmia*, among which the already-mentioned mason bee is generally reckoned, makes buildings before which it is impossible, as Blanchard says (*loc. cit.*), not to fall into ecstasy. It shows incredible prudence and reflection in the choice of materials for its cells. As it does not possess the shovel-like hollows in its hind-legs in which the honey bee collects pollen, it manages by rubbing its hairy body over the stamens, and on arriving at home brushes off the pollen, quantities of which remain between the hairs, from its body with its hind-legs. It has already been said that the mason bee closes its larval cells with a firm roof of mortar, and as this mortar becomes as hard as a stone in contact with the air it would seem inconceivable how the young bees should emerge if the wise builder did not leave a little hole closed only with soft earth or bits of stone, which look like the roof, in the near side of that cell the inmate of which will first emerge. It also very well understands how to accommodate its building to circumstances, and when it finds an old and deserted nest it will spare itself the trouble of making a new one, and adapt the old nest to its purposes after previous cleansing. In Algiers mason bees have also been observed which also shirk the labor, and make their cells in empty snail-shells. Others appear not to follow their inborn tendency to build, or building instinct, at all, but to usurp a ready-made nest with its cells at a moment when the maker is away, and then to hold it by force against the rightful owner. "Thus," adds Blanchard to the relation of this fact—which happens with nearly all nest-building insects (and animals) and has already been mentioned in the Introduction to this work—" single individuals of the same species seem to possess quite different tendencies. One is industrious and works honestly ; others are lazy and try to possess themselves of their neighbor's property by craft or force. Are there yet people who are ignorant enough to regard animals as machines, and to grasp nothing of the greatness of creation ? "

"Does the mason bee act like a machine," says E. Menault (*loc. cit.* p. 36), " when it directs its work according to circumstances, possesses itself of old nests, cleanses and improves them, and thereby shows that it can fully appreciate the immediate position ? Can one believe that no kind of reflexion is here necessary ? "

Bates noticed very similar facts (*loc. cit.*, Bk. ii., p. 43, etc.) among some species of South American *Melipona*, bees which have no sting, and yield a rather finer honey than their European relations. They utilise the baskets on their hind-legs not only for pollen, but also for collecting earth and carrying it home for building. They hang their combs in hollow trees or in holes in a bank, and require the earth to build up the opening of the hole, and to make only a narrow passage. They are thus masons and honey-gatherers at the same time. One small species is so prudent as to put a tube of kneaded earth, with the outer opening shaped like a trumpet, before its entrance, and here it keeps a continual guard, protecting the passage. Another kind collects leaves and chips, which it fastens together with resin from trees, to close its nest.

Dròry ("Eichstädter Bee Journal," 1874, No. 24) who kept some *Meliponæ* sent to him from Brazil for a year near his hive bees at Bordeaux, saw how these used accidentally spread varnish for building, and how individuals building tried to steal materials from each other. Cells for larvæ and for provisions are very different with them. Some species (as the already mentioned *Melipona scutellaris*) show great courage, and defend their nest very vigorously not only against other insects but also against men who try to destroy them. These wild bees are also perfectly competent to distinguish friend from foe among men, and to act accordingly. This was very strikingly shown in a case observed by Stedmann ("Travels in Surinam," Vol. ii., p. 286). Mr. Stedmann was visited in his hut by a neighbor, who had hardly entered when he rushed out again as if mad, yelling with pain, and ran to the nearest river to plunge his head into the water. It was soon seen that, being a very tall man, he had on entering the hut struck his head against a nest of wild bees which had settled in the roof over the door. Stedmann, distressed at such an occurrence, left the hut at once and bade the slaves destroy the nest. They were just going to obey when an old negro came up and declared that the bees would never sting Mr. Stedmann; he would undergo any punishment if they did. "Massa," said he, "you would have been stung long ago if you had been hostile to the bees. But as you are their host and have allowed them to build under your roof, they know you and your people, and will never do any harm either to them or to you." Mr. Stedmann found that the old man was right, for even when he shook the nest the bees stung neither him nor his negroes. The same old man related that he had once lived on a property on which there was a large tree. In this tree had lived, as long as he could remember, a society of birds and one of bees, both dwelling together in great amity. For if strange birds molested the bees they were driven away by those which were at home, while if strange bees came to the birds' nests they were attacked and killed by the friendly bees. The family of the owner of the property took great pleasure in this remarkable friendship and would not suffer it to be in any way disturbed.

Bates found, in the neighborhood of Santarem and Villa Nuova, no less than 140 different species of bees, which are mostly quite different from the European. Many build in hollow branches and boughs of trees, while others neither build nor gather stores, but lay their eggs in the nests of their companions. They are, therefore, among bees what the cuckoo is among birds. The habits of wild bees especially show the most manifold varieties according to varieties of circumstances, locality, etc. The wood bees of South America, instead of visiting flowers, collect the excretions of trees and the excrements of birds on leaves. In Abyssinia they sometimes build in the deserted dwellings of white ants, sometimes on the roofs of houses, sometimes in trees or rock crevices, etc., and always know how to choose the place of their settlement with regard to the best food. A bird lives at the Cape, the so-called honey-guide *(Cucculus indicator)*, which leads the honey-seeking Hottentots to the bee-nests known to it by flying before them, always to short distances, to show them the way. As a reward, therefore, it always receives part of the spoil. Can this also be instinct?

Very interesting also is the poppy or tapestry bee *(Apis, or Osmia, or Anthocopa papaveris)*, which digs holes for her larvæ three inches deep in the ground, and then so carefully lines or tapestries them with cut-out pieces of the soft and delicate petals of the wild poppy that not a wrinkle is left. In order to make the nest quite warm and firm, several layers of petals are laid one over the other. But most remarkable is the way in which, after egg and bee-bread have been placed in the cell, she closes it by fastening up the leaves just as we should tie a sack. This done, loose earth is piled over all, so that nothing betrays the presence of the nest.

There are a number of species of bees besides the tapestry bee which cut leaves with their long sharp mandibles, armed with four teeth, and are therefore called leaf-cutter bees by Réaumur. The most widely spread of these is the rose bee *(Megachile centuncularis)*, which cuts off pieces of rose and ash-leaves, and so arranges the cut-out pieces in her subterranean passages that a row of half overlapping covered thimble-shaped cups is formed, which serve as cells for the larvæ and their food. The arrangement and closure of each cell are as firm as they are neat. The whole is covered with earth, so as to be invisible from outside. Bingley very well describes *(loc. cit.,* vol. iv., p. 155) the care and thought, as well as mechanical skill, with which the leaves are cut out.

Réaumur relates a strikingly quaint anecdote of this insect:—

"In the early days of July, 1736, the lord of a village near Andelis came to the Abbé Nollet, in company with his apparently much terrified gardener. The latter had come to Paris to tell his master that his property had been bewitched. He had had the courage to bring with him some proofs which had convinced himself, the priest of the village, and all the neighbors of the witchcraft at work. The lord shewed these first to his doctor, who was able to give no opinion upon them, and then to M. Nollet, from whom M. Réaumur had the story. They were the cylindrical nests of the rose bee made out of rose leaves, which the gardener thought could only have been made by a man or a wizard. For, as no ordinary man could make them, and it was difficult to see with what object he should make them and bury them in the ground, they could only be the work of a wizard. M. Nollet assured the good man that they were the work of insects, and, as a proof, pulled a large maggot out of the leaves. When the gardener saw this, his erstwhile gloomy face brightened, and he looked like a man who had safely escaped a great danger."

The heavy carpenter bee (*Xylocopa*), spread in countless species all over the globe, understands very thoroughly how to hollow out her nests in old wood or beams. One of the commonest sort is our violet carpenter bee (*Xylocopa violacea*), whose toilsome buildings may easily be found by anyone. Solely by means of its strong sharp mandibles it hollows out with tireless patience admirably smooth cylinders in the wood, which it then divides into single cell-like chambers with sawdust fastened together with a kind of glue. From the lowest cell, in which is the larva which will be the first to emerge, the mother makes a passage to the outside, so that the insects lying above have to bore through the thin partition-walls of their respective cells in order to escape.

The carpenter bee requires no special lining for its cells, owing to the softness and dryness of the wood; but the wool bee *(Anthidium)* carefully lines or tapestries its cylinders, dug in loamy or sandy earth or in clay walls, with vegetable wool scraped off leaves and flowers. The cleverness with which it scrapes this wool off the plants is really astonishing.

The good-natured and rather limited humble bee, with its simple nest, displays unusual intelligence in biting the nectaries of flowers, as mentioned above, in order to be better able to reach their sweet contents. The way in which the moss humble bees surround their nest with a layer of wax, and then with a thick covering of moss, and in which they pass the moss to their abode, standing in rows and throwing a little bit of moss from one to another, betrays similar intelligence. The long passage to the nest, often winding, has, as a rule, a guard stationed in it, which has to drive away ants and other insects. Gödart also states (Brehm, *loc. cit.*, ix., p. 219) that each humble bee's nest has a trumpeter, who mounts to the roof early in the morning, flutters his wings, and wakes the inhabitants to work by trumpeting! The industry and intelligence of the humble bees increase, however, with the size of their community. Small societies confine themselves to what is necessary, prepare no wax covering, do not lengthen their honey-cells, etc., while in larger ones a kind of emulation in the best care of the house and tendance of the young impels them to increased efforts.

Each species of humble bee, like most species of bees, and like so many races of insects, has parasites or spongers resembling it in form and appearance, which utilise this likeness to put their eggs in the prepared nests, and then fly away without troubling themselves further about them. They have neither pleasure in nor ability for work, for their work instruments are rudimentary, apparently by long disuse; they lose nothing also thereby, for their eggs are brought to maturity with those of their host.

THE WASPS.

CHAPTER XXIV.

General Details.

INFERIOR in intelligence, but stronger or hastier in character than the bees, is the whole nearly allied large and warlike family of the wasps, which although far more simply organised have as numerous colonies as do bees, termites, and ants. The wasp State is arranged just like that of the humble bee. Since wasps and humble bees do not live over the winter, the alone surviving fertile female builds a nest in the spring underground or in some other favorable spot, lays eggs in it and nurses the larvæ hatched therefrom until the young emerge and are able to assist the mother in the further building and the tendance of the brood. The insects which emerge during this period are incapable of laying eggs although they are of the female sex; their whole activity is devoted to the business of nest-building and brood-tending, by which work their sexual organs are arrested. As with bees and ants, they are the neuters or workers. Toward the end of summer the female lays both eggs from which males are hatched and others from which fertile females are developed. These males and females unite in the autumn. As soon as the cold weather begins all save the fertile females die; the latter live through the winter and found new colonies in the following spring.

In these proceedings two things were for a long time regarded as inexplicable, and were considered as a mystery evidencing a very special and wise disposition of providence. The first puzzle, or the appearance of sexless workers among male and female animals was explained, as soon as it was understood that the so-called neuters—as already shown with bees and ants—were not really sexless, but were only undeveloped females, whose sexual organs did not develop owing to the expenditure of strength on nest-building and on the tendance of the young, while on the other hand, as we have seen with the bees, rest and plentiful food bring about their development. The second and more difficult problem, or the later appearance of males and fertile females, was solved when the discovery was made among bees that the queen was able to control the laying of male and female eggs, and how this was done. The fertile females of the wasps, therefore, lay female and fertilised eggs only so long as the stock of semen in the spermatophore lasts. When this store is exhausted in late summer or autumn, male insects must necessarily be formed. But of the female or fertilised eggs only the last laid will result in sexually-mature females, because the nest building is only then finished, and food enough can then be supplied to the workers to enable the sexual organs of the larvæ to attain full development. "That which at first appeared as a designed plan," says W. Dundt ("Lectures on Mind in Men and.Animals," ii., p. 196), "which mysteriously found its fulfilment by the instinct of animals, has been shown to be so completely the result of necessity in these most simple insect societies, that after the physical organisation of the animal has once been established in this distinct fashion no other explanation will be longer thought of."

The wasps proper, like the bees, live socially and in regular communities, in which labor is divided among the males, females, and workers or neuters, although not as completely as among bees. They would fill us with wonder and astonishment as to the reasoned and artistic building of their dwellings, the care of their young, the order prevailing in their societies, if we only had them, and not their yet more intellectual relations the bees, before our eyes. They are brave, patient, versatile and crafty, and as they are subject to common and daily observation as they fly about ceaselessly in autumn, a number of anecdotes are related about them which illustrate their sense and their cunning. As the wasps have not like the bees comfortable homes, or tolerably suitable holes in trees, sheds, etc., they usually build their nests and cells hanging, fastening them to the branch of a tree, eaves of a roof, etc., by one or more threads of twisted wood-fibres, and covering all with a hanging roof made of paper-like material. All the individual cells have their openings downwards, so that the larva must hang head downwards and must hold on to the cell-walls with their papillæ. This style of building protects the nest as much as possible against the vicissitudes of the weather, and especially against rain, which cannot penetrate into the cells. Each species of wasp has its own special plan of building, and works the rough material in its own way, so that there are countless kinds of wasps' nests, all of which, however, deserve admiration for their neatness and convenience of structure, although all these buildings are only intended for a summer's use. Most species of wasps scrape off with their mandibles the somewhat weather-worn outer surface of boards, fences, boughs, etc., and fasten together with saliva the wood-fibres thus obtained into a smooth mass, much resembling grey blotting-paper. They first knead together pellets out of this mass, and work it up further on arriving at home. If they can steal any real paper they use it to save themselves work. Dry leaves also occasionally serve their purpose. The cells themselves in which they bring up their young are sometimes almost cylindrical empty globes, sometimes hexagonal cells like those of bees, and are arranged in horizontal layers or combs, one under the other, joined together by a sort of supporting framework, or many pillars, leaving room enough between for free communication and unhindered access to the young. The cells for males and fertile females are larger and of rather different form from that of the workers' cells. The wasps do not require special store-chambers, for they bring in no honey, and kill and throw out any larvæ that remain when the cold of winter approaches and they can find no more food for them outside. Specially artistic is the nest of the hornet (*Vespa crabo*), often fifty centimetres high and from thirty to forty centimetres in diameter, surrounded by a thick covering of waved scales or layers. This strong and dreaded robber, the terror of all winged flower-suckers, goes about among them like the devil among poor souls, and even carries off large butterflies to its hungry waiting brood. It scrapes up young birch and ash-bark, to build therewith its grey paper-like cells and nests, and harms, therefore, a large number of young trees. But it also uses rotten wood for the same purpose, and if it finds hollow trees in which it can take up its abode, it makes itself at home and builds with little care.

How often have hornets' nests aroused the wonder of those which have seen or found them for the first time, and made them fancy that they had discovered a valuable prize!

The nests of the common wasps (*Vespa vulgaris*) are smaller, and are sometimes found above as well as underground, and from outside with their paper-like coverings look very much like a lump of coal. None the less they often contain the enormous number of ten thousand cells, after they have originally consisted of only from eight to ten, and have been constantly enlarged according to the needs of the increasing population. The entrance is found, as a rule, at the lower end of the pear-shaped hanging nest, and is watched day and night by the wasps, as among bees and ants, by a guard or sentinel, which warns the population within of any approaching danger. The males work in the interior of the nest just like the ordinary workers, but they seem to confine their labor chiefly to the cleansing of the nest, the carrying out of dead bodies, etc. They are fed, like the real females and the workers busied in the nest, by the out-flying wasps, which bring home animal food and fruits, and are robbers and murderers as bold as they are crafty. Like falcons, they pounce upon other insects, tear or bite off head, wings or legs, and carry home the quivering stump. Flies and bees suffer specially from them. In butchers' shops, after they have satisfied themselves, they will tear and carry off pieces of meat half as large as themselves. With soft fruits they suck themselves as full as possible, and divide the superfluity at home from mouth to mouth among comrades and larvæ. As soon as a worker laden in this way reaches the nest, it is at once surrounded on all sides and relieved of its booty. The larvæ are fed, like young birds, from mouth to mouth, and it is curious to see with what eagerness and speed the female hurries from cell to cell and gives each larva its share. As soon as a larva changes and leaves the cell as an imago, the cell is, as with the bees, most carefully cleaned and prepared for the reception of another egg.

Dr. Darwin ("Zoonomia," sec. xvi.) tells of a wasp which he observed trying to fly through the air with a large fly it had caught, after it had torn off its head and abdomen. The wind was against it, and the wings remaining on the stump formed so great an hindrance that the wasp, in order to avoid it, flew down to the ground, tore them off, and then flew off unshackled with its burden. This often-told and much wondered at story really contains nothing exceptional, nothing that would be beyond the intellectual powers of a wasp. And indeed similar observations have been repeatedly made. Herr H. Löwenfels, of Coburg, writes to the author under date of November 23, 1875 : " Walking on a sunny but windy autumn day I saw something hovering in the air and carried quickly by the wind in a slanting direction to the ground, which struck me as curious. It was not a leaf nor anything like one. Wont to pass carelessly by no natural phænomenon, however trivial, I followed the object to unriddle the puzzle and went to the spot whereon it had fallen. I here found a robber-wasp busied in lifting from the ground a large fly which it had apparently killed. It succeeded indeed in its attempt, but had scarcely raised its prey a few inches above the ground when the wind caught the wings of the dead fly and they began to act like a sail. The wasp was clearly unable to resist this action, and was blown a little distance in the direction of the wind, whereupon it let itself fall to the ground with its prize. It now made no more attempts to fly, but with eager industry pulled off with its teeth the fly's wings which hindered it in its object. When this was quite done it seized the fly, which was heavier than itself, and flew off with it untroubled on its journey through the air at a height of about five feet. I abstain from any reflexions on this accurately observed fact."

Herr Albert Schlüter, of Sisterdale (Kendall county), in Texas, saw a somewhat similar incident, and communicated it to the author as follows, under the date June 30, 1876: " In the last year of the civil war, in the spring of 1865, I sat angling, as almost daily, in the Podernales (?), five miles from Fredericksburg, in the shade of a small wood on the bank. In the sand near me a colony of ant-lions had settled, and I now and then pushed a passing insect into their funnels. Suddenly with piercing alarum a cicada of exceptional size fell from above between the funnels, and shook down and spoilt a number of them by its convulsive movements, while it continued its cries. Immediately after it came a pursuing hornet of the size and color of the German (we have here one twice as large, which easily carries off a full-grown tobacco-caterpillar), threw itself upon it and stung it, it seemed, to death, for the noise and movement at once ceased. The murderer walked over its prey, which was considerably larger than itself, grasped its body with its feet, spread out its wings and tried to fly away with it. Its strength was not sufficient, and after many efforts it gave up the attempt. Half a minute went by; sitting astride on the corpse and motionless—only the wings occasionally jerking—it seems to reflect, and indeed not in vain. A mulberry-tree stood close by, really only a trunk—for the top had been broken off, clearly by the last flood—of about ten or twelve feet high. The hornet saw this trunk, dragged its prey toilsomely to the foot of it, and then up to the top. Arrived thereat, it rested for a moment, grasped its victim firmly and flew off with it to the prairies. That which it was unable to raise off the ground it could now carry easily once high in the air."

Th. Meenan (" Proc. of the Acad. of Nat.," Philadelphia, Jan. 22, 1878) observed a very similar case with *Vespa maculata*. He saw one of these wasps try in vain to raise from the ground a grasshopper it had killed. When all its efforts proved to be in vain, it pulled its prey to a maple-tree, about thirty feet off, mounted it with its prize and flew away from it. "This," adds the writer, "was more than instinct. It was reflexion and judgment, and the judgment was proved to be correct."

Birds often find the same difficulty, it being very difficult often for them to rise from the ground, but very easy to fly from a height.

Dr. Ludwig Nagel, of Schmölle, writes: "On a business journey the writer saw an ichneumon-fly (*Ichneumon luteus*) laden with a large field-spider (*Aranea* or *Tegenaria agrestis*) coming obliquely across the footpath. The spider had already been killed by the bite and the sting of the fly. The latter, which had grasped the spider by the hinder part of the body with its jaws, struggled bravely on, pushing its victim before it. But the weight was too heavy and would go no further. So the fly turned round and went backwards, pulling the spider after it. Its nest was in a rather hilly, rising grass border. This reached, its march was rendered more difficult by the grass and the sloping ground, and it was often obliged to stop, and sometimes it slipped a little way back again. None the less it finished its journey and pulled its victim into its nest."

Herr Merkel, of Gumbinnen, writes as follows to the author under date of February 8, 1876 : " Six years ago I was tenant of the Railway Tavern,—(?) the—(?) Ostbahn,* and as I had there much time and leisure, I busied myself with seeking fossils, the railway excavations offering a fine field. As I was one day searching a hollow in the strata with this object, I saw a little grey wasp creeping over it, which was pulling along a caterpillar about an inch in length, holding it firmly with its jaws while it had three legs on either side the caterpillar. The latter appeared to be dead, for it did not move when its bearer let it go and went to a distance of about a man's pace. It was clearly looking for something, for it ran quickly up and down, until at last it stood still before a little hole in the ground, about the size of a lead pencil. It slipped into this hole, came out again, ran to the caterpillar, grasped it as before, pulled it to the hole, put one end over it, then went to the other end and lifted it up so high that the caterpillar fell in. But a piece of it stuck out, and this did not seem to suit the wasp, for it began to pull it out again, using its jaws and both fore-feet with wonderful skill. It laid the caterpillar down near the hole, went in again and brought out several little stones of the size of small peas. It then again let the caterpillar fall into the hole in the way already described. As this time there was none of it to be seen, the wasp went half in itself, emitted a slight hum (an expression of satisfaction. L. B.), came out again and began to scratch round the hole with its hind-legs until it was quite filled up. It then hovered round, examined its work very closely, and as it was to its satisfaction gave another hum and flew away."

Herr K. B. Zelinka, Railway Inspector of the South Austrian Railway, writes from Graz on December 23, 1875 : " In the year 1868, in the middle of the summer, my duties called me to the station of St. Lorenz (station of the railway from Marburg to Frangenfeste in the Tyrol), in the Drauthal. The hot July sun made it seem wise to spend the middle of the day in the shade thrown by good luck by a tree in front of the little inn, which stood in a very malarious position on the bank of the Radlbach at its junction with

---

* [As the queries are in the original, I leave the German name; it is apparently some district " Eastern Railway."—Tr.]

the Drau. Just as I was finishing my modest dinner (I was sitting in the open air before the house, beneath the leafy roof of the tree), I noticed the swift flight backwards and forwards of a common wasp. At the same time I observed on a spider-web, glittering in the sun, a fine specimen of the cross-spider, letting itself slowly down. When it had descended to about three metres from the ground, the wasp pounced on it with lightning-speed, and stung it in its thick abdomen. The spider raised itself a little way, but the wasp flew at it again, and inflicted a second sting. The spider now fell to the ground, and the wasp at once settled on it, and furiously tore the skin off its body. It then flew up and circled round the spider, struggling in the agonies of death. Whenever these struggles became more vigorous it fell again upon its victim, and only flew away from the place when the spider showed no further signs of life. I fancied that before the fight I saw, the wasp fell into the web of the spider and was attacked by it, and that when it escaped it took the vengeance I have related."

Whether the last idea of the writer is correct, may be left as there given, but the desire of vengeance is certainly a part of the passionate, quarrelsome, and choleric temperament of the wasp. According to the credible account of Ratzeburg, a boy once pushed a mushroom stipe into the entrance of a wasps' nest, on the Herrnkrug, near Magdeburg, so that the inhabitants could not get out. Two days later he went, accompanied by a relative, to the stopped-up wasps' nest, to see what had happened, and some dozens of wasps fell upon him and stung him so badly that he was made seriously ill. His companion was left untouched.

The common expression "to stir up a wasps' nest" shows how carefully people must guard themselves against this sharp and irritable insect. Even among themselves wasps depart very far from the peaceable character of the bees, and sometimes fight with each other in the bitterest way, and the males, although larger and stronger, take to rapid flight before the stings of their working sisters. They distinguish friend from foe among men just as well as do bees. The missionary Gueinzius, in Port Natal (in Brehm, *loc. cit.*, ix., p. 252), had allowed a native species of wasp to build its nest within the door-posts of his house, and in spite of frequent interferences with the nest, was only once stung by a young wasp, whereas no Caffre ventured to approach the door, much less to pass through it, because he would undoubtedly have been subjected to a general attack from the quarrelsome insects.

CHAPTER XXV.

VARIOUS SPECIES OF WASPS.

AMONG the many very distinct species of wasps, the *Polistes gallica*, or the French wasp, deserves special mention. It is not peculiar to France, but is found over the greater part of Europe, in Asia Minor up to Persia, and in North Africa as far as Egypt. Von Siebold, who investigated very carefully this species of wasp, has come to the (after all, not startling) conclusion that many of their actions are not the result of instinct, but of conscious reflexion ("Parthenogenesis of the Arthropoda," Leipsic, 1871). This is seen in their defence of their nest against ants, which are seized with a spring by the jaws and carried away as far as possible from the nest, or against strange wasps of their own species which steal their larvæ to feed their own young, against which they are sometimes obliged to call in the aid of the workers of their nest. Stranger wasps are recognised as such by touching them with the antennæ. The small and neat nests of the *Polistes*, generally hung on plants, are easy to watch, for they have no covering like those of other wasps; they are, therefore, turned with their covered or closed sides towards the west, so that wind and rain, which come generally from that quarter, may not penetrate. How very well this little creature knows how to otherwise adapt the building of its nests to circumstances, and to choose suitable places for its object, is shown by the observations of Rouget (*Mem. de l'Academie de Dijon*), who found nests of *Polistes* near Dijon in the interstices of tiles on wall-copings, which give to these positions great shelter and warmth, and in old glasses and cups in dustheaps.

The housekeeping of the *Polistes*, which have only males and two sorts of females (a larger and a less developed smaller), mirrors, according to Graber (*loc. cit.*, ii., p. 91), the original condition of the bees. The smaller females have already lost the best part of the prerogative of their sex, the reproduction of their kind. The same naturalist had an opportunity of witnessing an interesting incident, proving the great intelligence of the *Polistes*. He took away from a nest-building insect, during a short absence, the work it had commenced, and quickly fastened on another piece of a nest, three times as large. The architect, on its return, flew uneasily round it, sat down near it, considering the matter, doing nothing till the next day, and on the following day decided to take possession of the strange nest, and build on from it! It had, therefore, by reflexion, attained to a complete comprehension of the state of affairs, and, instead of giving itself the trouble of building a new nest, it sensibly took possession of the one offered to it.

The most artistic and also the largest nest is built by the Brazilian *Polybia liliacea*. De Saussure says that we may "regard its nest as one of the greatest wonders of insect architecture." Blanchard saw one of its nests, which was 110 centimetres long and 117 centimetres in circumference; it was still uncompleted above, but nevertheless contained many thousands of cells. A small American species of wasp *(Chartergus nidulans)*, Réaumur's paper wasp, on the other hand, builds little bag-shaped nests, the paper-like material of which is of such fineness and perfection that a Parisian paper merchant, to whom the paper was shown without being told its origin, was quite enthusiastic about it, saying that no Parisian merchant could make such paper, it must have been made in Orleans. In Guiana there is a perfectly black wasp (*Tatua Morio*), which also makes a beautiful nest. It consists of about eight or ten horizontal layers of cells, or combs, lying over each other and fastened round the bough of a tree: the whole is covered with a spindle-shaped covering, which seems as if composed of fine paper, made by an artist hand.

At Santarem, South America, Bates (*loc. cit.*, ii., p. 40, etc.,) observed a social, yellow and black wasp, (*Pelopœus fistularis*), which like our mason bee built its nest out of potter's clay. It rolls this with its jaws into little pellets, which it then carries away. Its nest looks like a purse two inches long, and is attached to a branch or other spot. Bates had the opportunity of closely observing the process of building. Each fresh lump of clay was brought by the wasp-architect with a kind of triumphant song, which, as it began to work, changed into a joyful busy hum. The little clay pellet was laid on the edge of the nest wall, and then spread along it with jaws and under-lip. The building was trodden down all round and smoothed with the feet. The completion of the whole occupied about a week. The interior of these nests is filled up with half-dead spiders, which the mothers bring for the food of the larvæ and reduce to this half-dead condition by stinging them, as do these wasps with all insects destined for the nourishment of their young.

Another species, also observed by Bates (*Trypoxylon*), builds its three-inch long nest in the form of a water bottle, and makes such a noise over it that when several together are building a house the neighborhood is in quite a commotion. They also emit very different notes when they fly in with their load or away, and when they are at work.

In Europe there are different species of unsocial wall wasps, which mostly belong to the *Odynerus* species, and dig the holes for their young several inches deep in old clay walls or firm sand banks. They also, in remarkable fashion, make outside long projecting and very artistically-built chimney-like tubes, which serve as protective entrances to the nests. An egg is laid inside with from ten to twelve half-dead caterpillars piled upon each other, on which the hatched larva feeds until it begins to spin. The mother knows exactly how many and what sized caterpillars it must bring for each larva, and seems always to select the same species for her young. Wesmael (in Brehm, *loc. cit.* ix., p. 240) relates that a mason wasp pinched between its jaws a caterpillar rolled up in a leaf until it left its sheltering cover and fell a prey to the hunter!

"In the first days of June," says Blanchard (*lot. cit.* p. 398), "we found ourselves, myself and two friends, in the Département du Nord, at a short distance from Denain, when our notice was called to an attractive spectacle. The road was bordered by a bank about two metres high, edging a large clover field. The bank was of firm soil and faced the full mid-day sun. Thousands, nay, hundreds of thousands of wasps were flying over the clover field, and picking little green caterpillars off the plants with indescribable eagerness. Others were busy digging holes in the earth, building chimneys, forming passages ; each single individual followed its own business with exemplary diligence, without troubling itself with its thousand co-laborers. No description could give an adequate picture of this lively and exciting scene. It was life itself in countless changing shapes. All these busy little creatures seemed to know their tasks, or at least acted as though they were fully conscious of the weighty duty they had to discharge in life. Is it not the same feeling, the same emotion, which is the motive power in every kind of society ? Each feels itself useful, indispensable, even in the lowest position. At the foot of the bank, where were the greatest number of insects, their edifices were seen in all stages of completion, for all the individuals had not come into the world at the same time. Some were making holes, others chimneys, others again were attending to the provisioning of their cells. The chimneys, as a rule three centimetres long, are slightly bent or curved on the side towards the ground, so that the rain cannot penetrate, and resemble lace (*dentelle faconnée*) made out of earthy material, a number of little interstices remaining between the circular cylinders or borders. They are therefore very brittle, while they are sufficiently strong for the insects themselves. When the egg is laid and the cell provisioned, this outer court is immediately destroyed, and the entrance closed up with the materials thereby obtained."

According to Perty (*loc. cit.*, p. 313), one of the mason wasps has been seen which put on its back a caterpillar trying to hold fast with its feet; just as was described earlier among the ants, and so carried it over the leaves to its nest.

Similar, but simpler than the mason wasp, and almost exactly as Herr Markel, of Gumbinnen, has described it, is the ever-active common sand wasp (*Ammophila sabulosa*), belonging to the large family of the solitary butcher wasps, or *Sphegidæ*. It digs a hole in sandy soil, puts in a captive caterpillar or spider, disabled by bite and sting, buries its victim when it has laid an egg thereupon, and covers over both. The larva, hatched in a few days, eats up the half-alive prey, and then spins its cocoon, from which, when the pupa stage is ended, it flies off as a wasp. In Bingley (*loc. cit.* iv., p. 139) is found a description of the whole proceeding, observed by a Mr. Ray, which agrees with that of Herr

Markel, almost word for word. The caterpillar put in was three times as large as the insect. The latter first pulled away a little ball of earth with which it had covered the opening of the hole, visited the interior, and then pulled in the caterpillar. The hole was then filled up with little stones and sand, and the earth smoothed over. At last the insect put two small pointed leaves on the spot where the opening had been, apparently, as the observer thought, so as to recognise the place again!* The North American blue sand wasp (*Sphex* or *Ammophila cyanea*, *Ichneumon cærulea*), described by Bates, acts just in the same way as is related by this writer of the *Pelopæus fistularis*. It builds cylindrical clay cells or clay tubes with divisions for its eggs, and fills them with captured insects, such as spiders, as food for the young when hatched. During building it emits a peculiar singing note, which can be heard at ten yards distance, and which seems to lighten its toil. It carries off spiders as large as itself, and, when they are too heavy to fly with, pulls them along the ground. Mr. Catesby weighed a wasp and a spider carried by it into the nest, and found the weight of the latter to be eight times as great as that of the former.

The grasshoppers, which the Pennsylvanian sand wasp (*Ammophila* or *Sphex Pennsylvanica*) carries into its holes as food for its young, are, as a rule, stronger and larger than the robber, which falls upon them suddenly from behind and disables them with its sting, so that they can make no resistance. As already mentioned, all the butcher wasps treat their prey in this fashion, and with the well-conceived design of making them defenceless but of not killing them, as they would otherwise soon decay in the nests and so be useless for the proposed object. There are also some digging wasps which, like *Bembex*, bring daily fresh nourishment to their young.

According to Taschenberg (in Brehm, *loc. cit.*, ix., p. 277), Gueinzius saw a butcher wasp (*Pompilus Natalensis*) which followed a large female spider through the open door into

---

* According to Taschenberg (Brehm, *loc. cit.*, ix., p. 283,) the sand wasp shuts off the entrance of its nest in a similar way, in order to make it impossible for parasitic insects flying about to lay their eggs therein.

the interior of his house, here disabled the poor creature after a desperate defence, and then, after a dance of triumph round its victim, carried it out of the door. According to the same author the hunting of the spider by the butcher wasps was already known to Aristotle.

The patience of the butcher wasps in attaining their object is wonderful. Fabre took away its prey from a *Sphex* wasp, which had carried away a slain grasshopper to its earth-hole, at the moment when it had crawled into the hole to visit it, and laid it down at a certain distance forty times over. But forty times it brought it back and went into the hole again on each occasion ere it prepared to pull it in (Brehm, *loc. cit.*, ix., p. 280). The prize is often stolen away by other wasps at this moment.

Specially interesting among the butcher wasps, owing to its striking behavior, is *Philanthus apivorus*, or the bee-eater, which hovers carelessly among the flowers as though it had nothing more to wish for. But those who watch it carefully will soon see an interesting spectacle. A bee appears; it is busily engaged gathering honey and pollen, and enthralled with its work gives no heed to its surroundings. The crafty *Philanthus* watches it closely, and when the opportunity appears favorable, darts at it with indescribable speed. It seizes it between head and thorax, and always succeeds in flinging it on its shoulders and piercing it with its sting. The bee naturally makes the most energetic resistance, but the *Philanthus* is more dexterous, and scarcely ever fails in securing its prey. After it is stung the bee writhes a brief space, tries to sting, stretches out its proboscis, and then falls down motionless. The murderer seizes it with jaws and feet and hurries away to its nest. Arrived thereat, it halts a moment, as though suspecting danger. It then picks up its prey again, brings it into the hole, lays an egg on it, closes up the place, and disappears. Its audacity is sometimes so great that it will even approach the beehive, although it is there menaced by the greatest peril, and invite open battle. Perhaps it is the *Crabo* of the old Romans, which, according to the description of the poets, fights *imparibus armis* (with unequal arms).

The habits of the ichneumon-flies, or *Ichneumonidœ*, are also deserving of notice; these seek the eggs, larvæ and pupæ of other insects in order to insert their own eggs therein by means of a long ovipositor, and thereby to obtain sufficient nourishment for the hatched grubs. They most frequently select the caterpillars of butterflies, which continue to live and to eat with the foreign guest in their bodies until the parasite injures some vital part and becomes a pupa. The mother, moreover, is not indifferent what young she chooses for her posterity; she also knows with marvellous skill how to snatch a suitable opportunity, however hidden it may be. Thomas Marsham (in Bingley, *loc. cit.* iv., p. 134) observed in June, 1787, an ichneumon-fly on a wooden post in Kensington Gardens. It was moving rapidly, holding its antennæ bent downwards. It felt about with these until it discovered the hole of an insect. When this was found it popped in its head and antennæ, and waited for about a minute in this position seeming very busy. The hole was then examined with equal care from the other side. The insect next turned round, measured the distance, and inserted into the hole the long ovipositor at the end of its abdomen. It remained in this situation for about two minutes, drew out the ovipositor, flew round the hole, and felt about in it again with its antennæ for about a minute. The ovipositor was then again introduced. The whole operation was repeated three times, one after another, before the eyes of the observer, but he approached too closely in order to see more exactly, and the insect flew away.

A week later Mr. Marsham saw several ichneumon-flies at work at the same place. They were apparently driving their ovipositors half their length into the firm wood, a thing that seemed impossible. But more exact investigation showed that the boring was made each time into the middle of a small white spot which consisted of fine white sand, and was a hole made by the *Apis maxillosa* (a species of bee) within which was a young bee larva. In very deep unclosed holes the creature sometimes crawled backwards so far that only its head, two fore-legs and wings, outstretched like arms, were visible. Mr. Marsham often saw the fly leave the hole again after investigating it, clearly because it had found it empty.

The fact that the solitary insects do not reach the high grade of intelligence and ingenuity evinced by their social relations, living in ordered communities, may easily be explained by the influence of the society itself, and by the highly-developed division of labor found in those States. It is not otherwise among men, and culture only attains its highest development when an ordered polity assigns to each his individual place, where the common life and the common work of many for a common object develops the slumbering powers and capacities of each, whereas these powers would remain ever in concealment among solitary, unsocial men. And have we not already seen in the humble bee that the manifestation of its intelligence and its industry appears in the measure in which its societies increase?

THE SPIDERS.

CHAPTER XXVI.

GENERAL DETAILS.

THERE is none the less among insects (in the wider sense), or among articulate animals, one class of creatures which in spite of their proverbial tendency to insulation, must, on account of their intellectual powers as well as of their ingenuity, be set by the side of the insects hitherto spoken of, while the same can be said of no other class or family, with the exception perhaps of some species of beetles. This class consists of the hated, feared and despised spiders, which only seem to exist in order that anyone who sees them may as quickly as possible destroy, drive away, or kill them. Those, however, who study their ways and doings will, in spite of their hideous exterior, feel themselves on nearer knowledge to be attracted rather than repelled.

" For all students," says Blanchard (*loc. cit.*, p. 669), " the spiders are the most interesting creatures of the animated world. In the most perfect representatives of this class we find narrowed into the smallest compass a richness of organisation which equals the greatest marvels of anatomy. The most remarkable instincts and often intelligence, are found therein, manifesting themselves in actions of the highest reason."

"The disposition and conduct of spiders," says Giebel (*loc. cit.*, p. 370) " claim the greatest interest, and in no way justify the common contempt and avoidance. Their movements are rapid, powerful and dexterous, their sensibility very great, their endurance, their courage on attack, their skill in the web, their tenacity of life, are wonderful. All their life-phænomena enthrall the attentive observer."

" Among all hunting animals," says Fée (*loc. cit.*, p. 104), " there is not one which can compare with spiders and with their skill in setting nets to catch their prey. Nor has any their patience and endurance."

" Looking at the animal scale from below upwards," says

Scheitlin (*loc. cit.*, p. 429), "we see the spider very far up, and we might almost fancy that a small or, indeed, any animal could not climb further."

Spiders have also received justice from teachers in the earliest times. King Solomon recommended them to his courtiers as a model of industry, skill, prudence, self-restraint and virtue; and Aristotle, as the oldest naturalist, gave them his full attention.

The web of the spider, which it spins in the most different places for catching its prey, has attracted most notice, and it has been regarded, like the cells of the bee, as the result of a peculiar inborn and instinctive artistic impulse. But far more even than the bee-cells does the spider-web differ and change with every variety of circumstance and position. Each species of spider or, we may say, each individual spider, follows its own peculiar plan in forming its web, and knows how to make it so as to suit the place, or to adapt it to circumstances. While the cross-spider spins the well-known and much admired wheel-shaped web and, hangs it perpendicularly, the sack-spider makes flat purse-shaped horizontally hanging nets, the threads of which run irregularly through each other and in the depth is a little bag for the reception of the maker. The famous *Malmignatto* of Corsica, Sardinia and a part of the Italian mainlands, only makes single threads from stones and posts haunted by large insects. Some build horizontal, some perpendicular nets. Garden-spiders spin threads from the ground to projecting stones, and catch in these no flying but only running and jumping insects. The species belonging to the genus *Scytodes* weave their strong horizontal webs behind into a narrow tube serving as a hiding-place, and spin radiating threads from its opening. Many species do not spin any webs, but prefer to catch their prey in a shorter fashion by springing and running. They only draw thread from their spinnerets, with which all true spiders are provided, for special purposes, as for instance for covering their eggs. The most dreaded amongst them are the tiger-spiders, which run about on walls, and creep slowly towards their victim in cat-like fashion, suddenly pouncing upon it from above with a powerful spring of as much as two inches. Others again, like the large bird-catching or hunting spider [*Mygale avicularia*. —Tr.] lurk in earth-holes, holes in branches, under stones, leaves, etc., watching for their prey, while still others, like the so-called trap-door spiders, of which we shall presently give more details, only go out hunting at night, and remain during the day in underground passages dug out by themselves, the opening closed by a cover which they can open and shut at pleasure. As spiders spin different kinds of webs, says an intelligent writer in *Chamber's Journal*, so also they live in all kinds of houses; and there is as much difference between these as between a Gothic tower and an Italian villa, between a Swiss *chalet* and a wigwam in the *Terra del Fuego*.

One of the most artistic nets is spun by the *Epeira basilica*, quite lately observed by the Rev. H. C. McCook, of Philadelphia, during his Texan ant excursions on the banks of the Colorado. Within a large pyramid-shaped net of irregularly interwoven threads, hangs an extremely neatly spun semispherical dome, from three to eight inches wide at the bottom, resembling to a certain extent the dome-shaped temples or basilicas of the early Christians, and thence the architect takes its name! "I have," says the discoverer, "never met with a prettier piece of work among the many spider's webs that I have seen and studied." From the centre of the dome hangs the gracefully-shaped, brilliantly colored, glittering builder, and watches for its prey. But what makes this spider particularly interesting is that it forms a perfect link or gradation between the wheel and the web-spiders—a fact more fully discussed by Mr. McCook in his article ("Proc. of the Aca. of Nat. of Philadelphia," April, 1878).

That all spiders use their threads for other purposes than weaving their webs—and before all things for making their cocoons, as well as for getting from place to place, descending from heights, for flight, for enveloping their victims, for lining their dwellings, for protection against the cold of winter, etc.—is so well known that it scarcely needs mention. It is less well known that the emerged young at first merely spin a very irregular web, and only gradually learn to make a larger and finer one, so that here again as everywhere else practice and experience play a great part.

Practice, experience, and reflexion must also guide the spider in the important choice of the locality in which it shall spin its nest, in order to catch the largest amount of prey. Before all it likes those places where the rays of the sun and dancing midges may be united with the possibility of a hidden retreat for itself, or where a slight draught blows flying insects into its outspread nest, or where ripe fruits attract them. The position must also offer favorable opposite points for the attachment of the web itself. People have often puzzled their brains, wondering how spiders, without being able to fly, had managed first to stretch their web through the air between two opposite points. But the little creature succeeds in accomplishing this difficult task in the most various and ingenious ways. It either, when the distance is not too great, throws a moist viscid pellet, joined to a thread, which will stick where it touches; or hangs itself by a thread in the air and lets itself be driven by the wind to the spot; or crawls there, letting out a thread as it goes, and then pulls it taut when arrived at the desired place; or floats a number of threads in the air and waits till the wind has thrown them here or there. The main or radial threads which fasten the web possess such a high degree of elasticity, that they tighten themselves between two distant points to which the spider has crawled, without it being necessary for the latter to pull them towards itself. When the little artist has once got a single thread at its disposition, it strengthens this until it is sufficiently strong for it to run backwards and forwards thereupon, and to spin therefrom the web. It behaves, therefore, exactly as did men when they spanned the terrible fall of Niagara with a chain-bridge. A paper-kite, such as children play with, was carried by the wind to the opposite side, and the strong cord which held it was used to draw over a heavier rope. The rope served the same purpose in its turn, and from this weak beginning the giant work was completed which now, like a spider's web, spans the flood and joins America to England.

The long main threads, with the help of which the spider begins and attaches its web, are always the thickest and strongest, while the others, forming the web itself, are considerably weaker. Injuries to the web at any spot the spider very quickly repairs, but without keeping to the original plan, and without taking more trouble than is absolutely necessary. Most spiders' webs, therefore, if closely looked into are found to be somewhat irregular. When a storm threatens, the spider, which is very economical with its valuable spinning material, spins no web, for it knows that the storm will tear it in pieces and waste its pains, and it also does not mend a web which has been torn. If it is seen spinning or mending, on the other hand, fine weather may generally be reckoned on, so that spiders have long served as weather prophets. Steady, fine weather is coming when the cross-spider weaves slowly and regularly; less good when it shows a certain amount of haste and as it were scamps its work. If it stretches a number of threads to strengthen its web, wind is to be looked for. The cross-spider in fine weather also sits in the middle of its web, while at night or in bad weather it draws back into a corner, whence it darts at its prey. If the latter be very large, so that the spider cannot master it or can only do so with difficulty, such as a bluebottle fly, a wasp, a bee, a grasshopper, etc., the robber approaches slowly and doubtfully, and generally prefers to let it escape, whereas smaller victims are at once spun over and so rendered defenceless. The spider has sometimes even been seen in the former cases to aid the escape of its unwelcome prisoner by biting through some of the threads. At other times, as Dr. Vinson observed with a Madagascan spider, it draws in front of or through the web proper a specially thick and strong thread, which is designed to hold such stronger insects or to prevent them from tearing the web.

It often happens that a widely spread web is not tightly enough stretched, and is more swayed backwards and forwards by the wind than is either convenient or useful to the spider. The sensible creature knows how to improve matters, and spins some strong threads from the ground upwards, attaching them to stones, plants, or other projecting points. This proceeding has indeed the disadvantage that men or animals passing beneath the web often tear the threads. But even then the spider manages to get out of the difficulty in a way which shows so high a grade of intelligence that one would hesitate to mention it, did it not depend on trustworthy observations. Gleditsch had already related how he saw a spider, in order to draw a web more tightly between two trees, let itself down by a thread to the ground, seize a small stone and then remount, fastening the stone to the lower part of its web so that people could pass comfortably below it. But a similar observation was made by Professor E. H. Weber, the famous anatomist and physiologist, and was published many years ago in Müller's Journal. A spider had stretched its web between two posts standing opposite each other, and had fastened it to a plant below for the third point. But as the attachment below was often broken by the garden work, by passers-by and in other ways, the little animal extricated itself from the difficulty by spinning its web round a little stone, and fastened this to the lower part of its web, swinging freely, and so to draw the web down by its weight instead of fastening it in this direction by a connecting thread. Carus ("Comp. Psychology, 1866," p. 76,) also made a similar observation. But the most interesting observation on this head is related by J. G. Wood ("Glimpses into Petland,") and repeated by Watson (*loc. cit.*, p. 455). One of my friends, says Wood, was accustomed to grant shelter to a number of garden spiders under a large verandah, and to watch their habits. One day a sharp storm broke out, and the wind raged so furiously through the garden that the spiders suffered damage from it, although sheltered by the verandah. The mainyards of one of these webs, as the sailors would call them, were broken, so that the web was blown hither and thither, like a slack sail in a storm. The spider made no fresh threads, but tried to help itself in another way. It let itself down to the ground by a thread and crawled to a place where lay some splintered pieces of a wooden fence thrown down by the storm. It fastened a thread to one of the bits of wood, turned back with it and hung it with a strong thread to the lower part of its nest, about five feet from the ground. The performance was a wonderful one, for the weight of the wood sufficed to keep the nest tolerably firm, while it was yet light enough to yield to the wind and so prevent further injury. The piece of wood was about two and a-half inches long and as thick as a goose-quill. On the following day a careless servant knocked her head against the wood and it fell down. But in the course of a few hours the spider had found it and brought it back to its place. When the storm ceased, the spider mended her web, broke the supporting thread in two, and let the wood fall to the ground!

The spiders are generally very careful about keeping their webs clean, partly because it then better serves its object, partly because it then rouses no suspicion of artifice among the victims swarming around. They therefore not only shake them from time to time so as to cleanse them from dust, but rid them directly of all larger rubbish accidently falling into them. Herr A. Frenzel, a smelting chemist, writes as follows to the author under date November 14th, 1875. from Freiberg, in Sáxony: " One day on rising from dinner I went into a room with a splinter in my hand, which I had used as a toothpick. A spider of the genus *Epeira* has spun her vertical web over a window in this room and sat quietly in the middle. I idly bit little bits off the splinter to bombard the spider with. I did not hit the spider but only the web, in which the bits remained hanging. When I had finished my bombardment the spider ran to the nearest bit, seized it, ran to the lower edge of the web, and let it fall to the ground. This manœuvre was repeated until every bit of wood had been cleared off the web. After a second attack on its abode with wood-splinters, the spider did not shrink from a second cleansing of its web." That in spite of their extreme shyness, spiders can be tamed, and become accustomed to human beings which show them kindnesses is proved by many facts and experiments which have attained to a certain amount of fame. Prisoners especially, in order to soften the rigors of solitude, have tamed spiders, so that they came at their call and took food from their hands, like the spiders of the unfortunate king Christian II. of Denmark. Dr. Moschkau, of Gohlis, near Leipsic, writes as follows to the author, on August 28th, 1876 : " In Oderwitz (?). where I lived in 1873 and 1874, I noticed one day in a half-dark corner of the ante-room a tolerably respectable spider's web, in which a well-fed cross-spider had made its home, and sat at the nest-opening early and late, watching for some flying or creeping food. I was accidentally several times a witness of the craft with which it caught its victim and rendered it harmless, and it soon became a regular duty to carry it flies several times during the day, which I laid down before its door with a pair of pincers. At first this feeding seemed to arouse small confidence, the pincers perhaps being in fault, for it let many of the flies escape again, or only seized them when it knew that they were within reach of its abode. After awhile, however, the spider came each time and took the flies out of the pincers and spun them over. The latter business was sometimes done so superficially, when I gave flies very quickly one after the other, that some of the already ensnared flies found time and opportunity to escape. This game was carried on by me for some weeks, as it seemed to me curious. But one day when the spider seemed very ravenous, and regularly flew at each fly offered to it, I began teasing it. As soon as it had got hold of the fly I pulled it back again with the pincers. It took this exceedingly ill. The first time, as I finally left the fly with it, it managed to forgive me, but when I later took a fly right away, our friendship was destroyed for ever. On the following day it treated my offered flies with contempt and would not move, and on the third day it had disappeared."

This shows that a spider can be both hurt and offended. And, indeed, its little mind is not incapable of the feeling of revenge. At least in Marquart's "*Les facultés intérieures*," p. 163, there is a communication from Reklus, according to which a spider inflicted a very poisonous bite on the forehead of a young man, who for several days, one after the other, destroyed her web spun each day in a very favorable position over a hole in a roof.

More certain than this is the remarkable love of spiders for music, which has been established by trustworthy and numerous observations. Spiders living in a room are attracted out by the playing of a piano, guitar, or violin, especially when the music is tender and not too loud. They approach as close as possible to the instrument and to the player, and seem so bewitched by it that they pay no attention to anything else. They are often seen to let themselves down by a thread from the ceiling of the room, so as to get as near as possible to the player. As soon, however, as the music becomes noisy they run back to their nest. Professor C. Reclam ("Body and Mind," 1859, p. 275), at a concert at Leipsic saw a spider which let itself down from one of the chandeliers during the playing of a violin solo, whereas each time that the orchestra broke in it ran swiftly back to its hiding-place. Similar observations have been published by Rabigot, Simonius, von Hartmann and others.

Spiders also know how to feign themselves dead, like so

many other insects, when their life is in peril, and manifest therein a real heroic indifference. "I have," says Smellie (in Bingley, *loc. cit.* iv., p. 232), " run needles through spiders under these circumstances, and torn them in pieces, without their betraying the slightest symptom of pain."

CHAPTER XXVII.

Various Species of Spiders.

ONE of the most interesting of the species of spiders is the *Argyroneta aquatica*, a swimmer-spider which should be regarded as the discoverer of the diving-bell. This remarkable creature lives with us in almost all stagnant pools, and stays under water for hours, although, like every other spider, it would be drowned by the entrance of water into its air-sacs, if it did not know how to prevent it. It raises its abdomen above the surface of the water, surrounds it with an air-bubble and dives down, the air being held apparently by the downy coating of its body, and looking like a glittering ball of silver or mercury. Arrived below, it seeks a place where water-plants grow thickly together, and rubs its legs against its abdomen till it loosens the air bubble, which is held down by the tangle of plants. This done it rises again to the surface, and repeats the proceeding until it has collected a good quantity of air at the same spot. It then surrounds this air with a very fine but thick web, which has perfectly the shape of a diving-bell, and is attached all round by stretched out threads. If the bell is not quite full of air, new bubbles are brought from the surface in the same way and emptied into it, and now, complete, it looks like a shining silver bell. In this poetical abode, reminding one of the "Arabian Nights' Tales," the insect lives, carries there its food and brings up its young. It hunts not only in the water but also on dry land, but always takes its prey to its hidden crystal palace. If any store of food remains after its hunger is satisfied, it is carefully attached to the diving-bell by threads. The male builds its light dwelling close by that of the female, and unites them by an opening or covered gallery. So live the wedded pair, each in its own house, in peaceful unity side by side, busied only with the care of their family, far from the bustle of the world, but yet ever supplied thence with the slightly softened beams of dazzling light. Happy spider couple!

In captivity the clever insect so thoroughly adapts itself to circumstances that it either hangs its bell to the side of the vessel, or if plants fail, draws some threads crosswise through the water, and hangs the bell between them. Our artist gladly passes the winter in an empty snailshell, closing the opening with an ingenious web.

Less idyllic than the water-spider is our native hunting-spider (*Dolomedes fimbriata*), which belongs to those species which spin no web, but hunt their victims like animals of prey. As the *Argyroneta* is the discoverer of the diving-bell, so may this be regarded as the discoverer or first builder of a floating raft. It is not content with hunting insects on land, but follows them on the water, on the surface of which it runs about with ease. It, however, needs a place to rest on, and makes it by rolling together dry leaves and such-like bodies, binding them into a firm whole with its silken threads. On this raft-like vessel it floats at the mercy of wind and waves, and if an unlucky water-insect comes for an instant to the surface of the water to breathe, the spider darts at it with lightning speed and carries it back to its raft to devour at its ease. Thus everywhere in nature are battle, craft and ingenuity, all following the merciless law of egoism, in order to maintain their own lives and to destroy those of others!

The largest and most dreaded of all the spiders is the tropical bird-catching or hunting-spider (*Mygale avicularia*), belonging to the family of the tube-spinners (*Tubitelæ*). Its strong powerful mandibular palpi rise threateningly from the upper edge of the maxillæ, and, with the help of these, it is able not only to master large insects, but also lizards, and even small birds. The latter fact has been doubted, but has been anew confirmed by Bates from his personal observation (*loc. cit.* i., p. 160). He saw a *Mygale* near the Amazon River which, with outstretched legs, measured seven, and, without them, two, inches long. Body and legs were covered with strong grey and red hair. Bates' attention was drawn to the hateful monster by a movement he noticed on a tree-trunk. He looked more closely at a deep crevice in the tree, which was covered by a thick white web. The lower part of the web was broken, and two small birds of the finch tribe were covered with the threads. They were about the size of the English siskin, and Bates thought they were cock and hen. One bird was quite dead; the other lay half-alive under the spider, smeared with the monster's filthy saliva. Bates drove away the spider and rescued the bird, which, however, died immediately.

The *Mygale* species are, as Bates adds, very numerous in Brazil. Some build under stones, others make tunnels in earth, and others, again, make holes in the thatched roofs of houses. The natives call them *Aranhas carangueijares*, or crab-spiders. The hairs with which they are covered remain sticking in the skin if they are touched, and cause a very painful irritation. Many are of enormous size. Bates one day saw some children which had fastened a string round the body of a *Mygale*, and pulled it behind them like a dog. Near Para, on the mouth of the Amazon, the *Mygales* are very numerous in sandy places, and exhibit the most various habits. Many build on or in houses nests or places of refuge of a fine close web much resembling fine muslin. Others build similar nests in trees, and it is these which catch birds. The *Mygale blondii*, a red-brown, hair-covered monster, five inches long, digs a tunnel in the earth about two feet long and two inches in diameter, the inner wall of which it tapestries with a handsome silver-like shining web. It only goes hunting at night, and shortly before sunset it may be seen watching at the mouth of its hole, swiftly disappearing within if it hears a heavy footfall in its neighborhood. Passing insects fall victims to its murderous bite.

Almost exactly in the same fashion behaves the mining, or as Moggridge named it, the trap-door spider (*Mygale* or *Cteniza cœmentaria* and *fodiens*, first designated by Moggridge as the *Nemesia cœmentaria*), an inhabitant of the South of Europe, also belonging to the family of the tube-spiders or, more exactly, to the *Territelariæ* (earth-workers); this animal, on account of the skill and the ingenuity with which it builds its subterranean dwelling, and protects it against attack from without, carries away the palm unchallenged among all species of spiders with regard to interest and intelligence, although it is far behind its Brazilian relations in bodily size. The mandibular palpi of the *Mygale fodiens* are armed with a kind of sharp rake, while its feet carry teeth like a comb. With the help of these instruments the animal digs out subterranean tunnels or passages, in which it can retire or hide as it pleases. The interior of this dwelling is lined in carefullest fashion with a fine silken web. At the entrance a door is made which, to fitly describe, as Blanchard said, all expressions of wonder would be insufficient. It has the shape of a lid, and consists of pieces of earth bound together with silken material. It is very thick, and is broader above than below, so as to close the hole as completely as possible. Its outer surface has exactly the appearance and condition of the surrounding ground, so that nothing betrays its presence, while the inner surface, like the rest of the dwelling, is lined with silken web. But this is not enough; the door, like all proper doors, has lock and hinge. The hinge is made of a very thick and strong silk; the lock consists of a row of small holes, into which the spider dwelling within inserts its claws, in order to hold the door fast from inside in case of approaching danger. When it goes out to hunt at night it lifts up the door and lets it fall behind it, just as do dwellers in human holes or cellar-dwellings. On its return it pulls the door up with its feet, and so glides into its subterranean abode.

The habits of this remarkable creature have been most exactly studied and described by J. T. Moggridge—who also so admirably observed the harvesting ants—in his already often-named book, "Harvesting Ants and Trap-door Spiders." According to him, the animal first became known in the second half of the last century by thorough observers like P. Browne, Sauvages and Rossi, and is therefore, with regard to antiquity and fame hallowed by ages, far behind ants and bees.

Without doubt, says Moggridge, are the webs and cylinders of common spiders very wonderful pieces of art, but in comparison with the works of the trap-door spiders they are no more than is a common tunnel, for instance, as compared with the tunnel of Mont Cenis. It is enchanting to see with what patience and skill the little creature, which must be reckoned one of the best natural artificers and discoverers, overcomes all difficulties and dangers.

Until Moggridge—who was inspired and led to his studies

by the Hon. Mrs. Richard Boyle, and who assigns to her the priority of discovery—only the two simplest forms of nest of the trap-door spider were known, while he has described several further forms which, by their complexity, not only prove the extraordinary ingenuity of the creature, but also seem to confirm the proposition long established by other experiments, and so overwhelmingly important in judging the intellectual capacities of animals, that progress and perfection are not the inheritance of man alone, but also, though in smaller measure, are to be met with among animals.

Of the two simple, or simplest forms of nest, the one which Moggridge calls the single door cork nest, has already been practically described. The other, which he names the single door wafer nest, is only found in the West Indies, and has merely a thin oblate-shaped door, made of silken web without mixture of earth, which lies loose over the mouth of the nest, without, as in the other simple form, entering like a cork into the opening of the cylinder.

The West Indian nests are far stronger and tougher than are the European, and have also a somewhat different form, giving them some likeness to a stocking. They are the work of the *Cteniza nidulans*, and Mr. P. H. Gosse, who has admirably described them, shows that there is the greatest difference in the amount of finish given to them. They are all, however, furnished within with soft delicate silvery silken material.

The other simple, or cork nest, differs at the first glance from the one above described by the much greater thickness of the door, and the therefore different fashion of closure. But between these two forms there is another, described by Professor Westwood ("Trans. of the Entom. Soc. London," 1841-43), a transitional form, built by the *Cteniza ædificatorius*. According to Moggridge, also, very nearly allied species sometimes build very unlike nests, while very distinct species often build very like or almost similar ones—a fact exceedingly against the instinct theory. This fact, however, is all the more surprising as all the working-instruments, especially the claw-like endings of the feet, seem to be very differently shaped in species far apart from each other, the converse being also true.

The nests are often very hard to find, for they are generally in damp shady places or shelving banks, where rubbish, rolled down earth; or rank vegetation covers them as much as possible. Sloping banks are also generally chosen, so that the doors may fall too by their own weight without difficulty, while the nests are very seldom found in flat ground. The door, as a rule, shuts very firmly and securely, although Moggridge, who investigated a large number of nests, here also found great individual differences between members of the same species. Cleverness is, therefore, a talent which is as differently divided among spiders as among men.

When Moggridge moved such a door with the point of a penknife, it was at once slowly drawn downwards, like the shutting down of the shell of a limpet. He then tried to open the door, in spite of the vigorous counter-efforts of the inhabitant, and saw how the spider held on with all its might, lying on its back and having hooked its feet fast into the silken covering of the lower part of the door. Moggridge did not force any wider opening, but cut out of the ground the upper part of the nest with the spider. It was then seen that the little holes on the inner part of the door, which the spider used for a hold for his feet, were only found on the part of the door away from the hinge, thus exactly where they were wanted with the object of holding fast. These holes, further, were not present in any other nests.

A nest found accidentally in digging up a plant which had been brought to Mr. Moggridge was quite covered on the surface with moss, and the moss grew on the surface of the door itself, and looked exactly like that growing all round. The deception was so complete that Mr. Moggridge was unable to distinguish the closed door, even when he held the nest in his hand. This is the more remarkable, as from all appearances it must be thought that the spider itself planted the moss on the surface of its door!

According to the descriptions given by Moggridge of the newly-discovered and more complicated nests, there is found in these on the surface of the ground a thin oblate-shaped door, resembling the West Indian form, and, from two to four inches deeper, a second and more strongly or solidly-built door, which last also is very differently shaped, accordingly as it belongs to a nest without or with ramifications.

The latter, or ramified form, is the one most frequently met with at Mentone. The main cylinder runs backwards into the earth, either straight or spirally, while the other, or secondary cylinders, bend away upwards at a sharp angle, and either, as is generally the case, end blindly, or, in rare cases, open on the surface of the ground. With the latter arrangement one of the two doors is generally neglected, and the upper part of the cylinder is half stopped up with earth, so that we are forced to conclude that, by some unfortunate mischance, the older door was injured, and the spider prepared a new one instead at another place. The secondary cylinders generally end like blind alleys, as has been said, and Moggridge found that this was always the case with very young spiders.

In these ramified nests with double doors, the upper door, held by the hinge and its own weight, only lies on the mouth of the nest, and does not enter the passage as in the single cork nest. The lower door also hangs by a hinge from the edge of the corner made by the division into two cylinders, and can be turned either way, so as to close the entrance either of the main or of the secondary cylinder. It is from a line to a line and a half thick, of elliptical form, ribbed above, smooth below, and with a flap at its lower end. The whole is made of earth, held together by silken web. When the door hangs so as to close the entrance of the secondary cylinder, it suits its surroundings so well in shape and appearance that it only looks like an unbroken part of the wall of the main passage.

If the upper part of such a nest be destroyed, the lower door will be seen to be secretly moved and the main cylinder closed off, being evidently pushed by the spider itself from above; the creature can even sometimes be caught in this situation, its shoulders pushing against the door. But when the spider sees that resistance is useless, it will either hide itself at the end of its cylinder rolled up like a ball, or rush out and strike with its fangs at the disturber of its peace.

The spider may behave somewhat differently when it has to defend itself against its natural enemies, such as ichneumon flies, sand wasps, ants, millipedes, little lizards, etc. It apparently shuts the subterranean door in the main cylinder first against the assailant, and draws back into the secondary cylinder when the other passage is forced, drawing the door after it. The intruder then finds nothing in the main cylinder, and, owing to the similar appearance of the door to the rest of the interior, is unable to discover the presence of the secondary passage!

As the outer door, owing to its slight thickness, cannot be clothed with vegetation, like the cork door, the spider endeavors to supply this want by weaving into it as much as possible veiling or deceptive material, such as dry leaves, splinters of wood, roots, grass-stalks, etc., so that it may harmonise with its surroundings. Yet there are some nests in which this manœuvre is so badly carried out that attention is rather attracted than avoided. Nests which lie on bare ground easily betray themselves by the fact that the doors dry more quickly and more easily than the earth around, and thereby take a lighter color. Leaves act as the best covering, and a single leaf often amply suffices.

Sometimes the cylinder rises like a chimney between moss, grass, stones, plants, etc., as much as two or three inches above the surface of the ground, and is attached all round by woven silken threads. Such above-ground cylinders were built by a spider (*Atypus piceus*) observed in the neighborhood of Paris, but are far more uncommon than the earth tunnels described, and have no doors.

The second form of nest newly discovered or newly described by Moggridge is again an unramified cylinder, and is the work of *Nemesia Eleanora*. The second or subterranean door lies from one to four inches deeper than the upper, and only serves to close the one somewhat narrowed cylinder. The upper door is for purposes of concealment, the second for purposes of defence. The latter is again made of earth and silken web, is from one to two lines thick, and has in the place of a lock a flap like that of the subterranean door of the ramified nest. This flap may serve to lay hold of, and by means of it the door pushed closely into the cylinder at the approach of a foe may again be drawn back when the danger is past. The door itself, as also is that of the ramified nest, is slightly hollowed above and slightly convex below, so that when open it may not interfere with free passage through the cylinder. It is somewhat less wide at the upper end than at the lower, so that the somewhat narrowing tube can be better stopped up, just as this is the case in reverse direction of the cork door. It is not as long as the double door of the ramified nest, which has to fulfil a double duty, but is rather broader and closer. All the doors are more or less elliptical, this being necessarily the case as they have to close the cylindrical in a somewhat sloping direction. Yet their outlines are occasionally different, according to the differences in the circumferences of the cylinders.

Moggridge several times found among the *N. Eleanora* a larger or smaller number of young ones with the mother in the cylinder, while this never chanced with the other species. He never saw the trap-door spiders, which go out hunting at night, out of the nest during the day, although other observers are said to have seen them.

Erber relates the following ("Verh. d. k. k. Zool. Bot., Gesellsch of Vienna," Bk. XVIII., pp. 905 and 906) of the *Cteniza ariana* (cover-spider), found in the island of Tinos in the Greek Archipelago, which he observed by moonlight: "The doors opened soon after nine o'clock; the spiders came out, fastened the open doors by a few threads to the surrounding grass or small stones, each spun a web of about six inches long and half-an-inch high, and went back into their dwellings. I had selected my position so as to be able to watch three spiders at the same time. They soon had some night beetles caught in their webs, and the spiders seized them at once. They sucked out their juices and pulled their dead bodies some feet away from their holes. On the following morning I again visited the place and found that the nets stretched during the night had wholly vanished. The door of the nest of one of the spiders, which I had captured during the night, was standing open, and I could clearly recognise the threads ornamented with dewdrops with which it had been fastened to the ground."

On the other hand, according to the information sent to Herr Hansard by a friend, there is a trap-door spider in the island of Formosa, which builds nests like those of *Cteniza fodiens*, and which is generally seen out of its nest during the day, but as soon as anyone goes near it, it darts into its dwelling, shutting the door behind it. According to Lady Parker these spiders are so common in Australia that scarcely any notice is taken of them. They also are outside in the daytime, and run into their nests when frightened. The doors shut so perfectly that they are very difficult to find.

On the other hand, the French trap-door spiders described by de Walkenaar work or hunt only by night. They also spin webs in the neighborhood of their nests. Costa says of the South Italian *Nemesia meridionalis* that it makes its nest very variously, according to the nature of the ground in which it builds, and that it makes the silken covering the thicker the looser the earth is. In very firm ground the cylinder, except close to the entrance, is smoothed and smeared over, while in other cases the creature builds so strong a tunnel that even when the earth is removed it stands open, the architect having had the prudence to attach it all round to separate steady points.

Moggridge watched the process of building of a captive spider, after he had previously made a cylindrical hole in the earth. But doors made in such cases are not generally as perfect as the natural ones, as though under the unaccustomed conditions the spider did not think it worth while to develop its whole art. They often build none at all, or, against their usual habit, make long closed webs of spun silk, which they stretch between the earth and the gauze covering their prison and in which they hide. These webs have some resemblance to the already described above-ground tubes made by the *Atypus*.

All the trap-door spiders kept by Moggridge were busier during the night than during the day. They have less apparently to fear during the night from their enemies (baboons, squirrels, birds, lizards, turtles, frogs, toads, wasps, etc.), while their game, such as ants, beetles, earwigs, woodlice, etc., are as much about during the night as they are themselves. All trap-door spiders do not spin webs, like those described by Erber, but many lie in wait at the entrance of their dwelling for passing insects, which they catch with a swift spring or grab and pull into their hole, pulling to the half open door behind them or letting it fall.

If the doors of the trap-door spiders' nests are pulled off, it is found that they are replaced anew. But if they are fastened down firmly with a pin, so that they cannot be opened, the inhabitant makes a fresh opening close by, and provides it with a door. This accident may also often happen naturally by chance events.

Moggridge one day tied open firmly with a thread three trap-doors standing in a row, so that they could not be shut again by the spiders. On the following day, however, he found one of the holes again partly closed, the silken lining being drawn across the opening of the passage from within. The second cylinder was unchanged. The third was temporarily covered by three olive-leaves, spun together and fastened by threads to the edges of the opening of the passage. Two days later a perfect movable door was made of them. Thus each of the three spiders had acted quite differently, each according to its own view and thought. How could this be instinct?

The trap-door spiders very unwillingly leave nests which they have once made. It has sometimes been observed that nests turned completely upside down by the digging of the ground are not left by their inmates, but are made habitable again by carrying the tunnel to the surface of the ground and the formation of a new door. In all cases no small time and trouble are devoted to the formation of these remarkable dwellings. The individual nests also only attain gradually their full perfection and size, having at first, while the spider is young and small, only the thickness of a crow-quill, but being continually enlarged with the growth of the spider itself. The door also is increased bit by bit, and thereby sometimes acquires the appearance of an oyster-shell. On close investigation it is found to consist generally of a large number of single layers of silk laid one over the other, with earth between, and which are sometimes as many as twenty or thirty. Sometimes, also, the old and now too small doors are left and new ones made, so that we may come across two or three of different sizes. It is obviously far more convenient and more advantageous for the growing spider to continually enlarge and increase its nest, than to make new nests time after time, for Moggridge has noticed differences of size of from one to sixteen lines of diameter, and the number of transitional nests would therefore have to be very large. How great must be the growth of spiders is seen from an account of F. Pollock ("Ann. and Mag. of Nat. Hist." for June, 1865), wherein a female *Epeira* or wheel-spider, after a lapse of eight months and ten or more changes of skin, was 2,700 times as heavy as at her birth.

The most common of the described forms of nests is that of the simple non-ramified single door cork nest. There are six different species, belonging to at least three genera, which make this kind of nest, whereas the other more complicated forms are in each case only built by a single species. Three of the accurately known species of the *Territelariæ*, or earth spiders (*Atypus piceus*, *A. Blackwallii* and *Nemesia cellicola*), build the simplest kind, the silken-lined tunnel without any covering at the mouth. As, however, Professor Ausserer, in a monograph on the earth spiders, has reckoned up no less than 215 species, a very wide field here remains open for discovery, and it is most highly probable that there are still many among them which build nests of the single door cork or of some other type.

Moggridge has seen Australian spiders' trap-doors of the type of from one to two inches in diameter, and considers that this type is spread over almost all the world, whereas his other types have only yet been found in certain limited spots. He adds, indeed, that this latter fact may change with time, but does not think that so wide a diffusion of them as of the simpler type will ever be shown. Let this be as it may, the architecture of the cylinder-building spiders shows a series of gradations and transitions from imperfection to greater perfection, as would be expected from the universal principles of the descent and development theory, and as is incompatible with the once for all determined rules or forms of behavior and building-tendency laid down for animals, according to the opinion of the instinct philosophers.

That the number of these transitions through the described forms is not yet nearly exhausted, is not only a supposition, but is actually proved by the issue of the additions found published by Mr. Moggridge in the form of a supplement or appendix, the year after the appearance of his interesting work. We learn therein of no less than three or four more new types of trap-door nests, hitherto unknown in Europe, so that the whole number of these types, apart from the still very imperfectly known but also very noteworthy *Atypus* species, mounts to six or seven. This variety of forms is much conditioned by the differences of the species building alike, for, as already said, the same species often build very different nests, while the same sort of nests belongs to the most different species, but depends far more, as might naturally be expected, on the variety of outside circumstances and life conditions. In

California the nests are made shallow and seldom exceed three inches in length, although the spiders that build them are very large and are also much dreaded owing to their poisonous bite. They leave their nest during the day, but slip back again at the slightest sign of danger. Their most dreaded foe is a large species of wasp, living in California. But little *Hymenopterœ*, which lay their eggs among those of the spider, are also dangerous to it or at least to its young, and it is therefore easy to see how useful and even necessary it is to the spider to have a strong and well-fitting door against these as well as against larger foes.

Several nests which Moggridge received from Palestine were very short, and had the greatest resemblance to those of the *Cteniza Moggridgii* of Mentone.

Near Bordeaux Moggridge found a number of non-ramified nests, which had not cork but oblate doors, yet nevertheless had no second door in the interior of the cylinder. The secondary cylinder here, turning away at a sharp angle from the main one, ran almost to the surface of the ground, where they are closed with earth and web in a way which is easy to penetrate in case of need. This plan appeared to yield the same advantage as the cork nests in their heavy door. Perhaps, as Moggridge suggests, with great probability, the simpler nest forms belong generally more to the colder, the more complex to the hotter climates.

Moggridge found near Hyères, in France, a ramified nest with double doors, the second or subterranean door of which is constructed in a quite peculiar way, with strong wedge-shaped projections on both sides and with a long flap. It lies tolerably high above, shuts very firmly and, when the outer door is open, is pressed into the main cylinder with great force from above and from the side tunnel.

Moggridge afterwards discovered in the already described ramified and double-doored nest with one oblate door, a further and remarkable complication which he had at first overlooked. This nest is the work of the *Nemesia Manderstjernæ*, and in addition to the upward-tending possesses also a side cylinder or side passage, so that this complication places the nest at the head of all, and proves its architect to be the most distinguished or cleverest among the tunnel-making earth-spiders. The most remarkable thing, however, about this arrangement is that any enemy penetrating into the nest is most completely deceived, for when the subterranean door is so situated that it shuts off the main passage, the invader in pressing onwards is not in the latter but in the side cylinder, and as he discovers nothing here, but imagines himself to be in the main cylinder, he must retire with all his work for nothing.

But that which complicates everything to a specially high degree is the circumstance that this side tunnel, as a rule, is only found in the nests of the younger spiders, while in the older ones it is generally stopped up with earth and rubbish, and is, therefore, easily overlooked in investigating the nest, or else is not regarded as being present. This curious state of affairs is explained either by the fact that the side cylinder is intended as a protection against an enemy which the older and stronger spiders have no need to fear, or that the older spiders neglect this precaution when the time is come that they lay no more eggs and, therefore, no longer need any special protection for them. As Moggridge, further, found many remains of slain insects in the doubtful cylinder of old nests, it is also possible that it is intended to serve as a place of deposit for these remnants.

The nests of *N. Manderstjernæ*, for the rest, exactly resemble those which Moggridge investigated at Mentone and described.

To show, lastly, how various are the transitional forms and gradations so important in deciding upon the gradual origin of the forms of nests, Moggridge also alludes to the similar buildings made by other genera of spiders. *Lycosa narbonensis*, a spider of Southern France much resembling the Apuleian tarantula, and belonging to the family of the wolf spiders, makes cylindrical holes in the earth, about one inch wide and three or four inches deep, in a perpendicular direction; when they have attained this depth they run further horizontally, and end in a three-cornered room, from one to two inches broad, the floor of which is covered with the remnants of dead insects. The whole nest is lined within with a thick silken material, and has at its opening—closed by no door—an above ground chimney-shaped extension, made of leaves, needles, moss, wood, etc., woven together with spider threads. These chimneys show various differences in their manner of building, and are intended chiefly, according to Moggridge, to prevent the sand blown about by the violent sea-winds from penetrating into the nests. During winter the opening is wholly and continuously woven over, and it is very well possible, or probable, that the process of reopening such a warm covering in the spring, after this opening was three quarters completed and was large enough to let the spider pass out, may have long ago awaked in the brain of some species of spider the idea of making a permanent and movable door. But from this to the practical construction of so perfect a door as we have learned to know, and even to the building of the exceedingly complicated nest of the *N. Manderstjernæ*, through all the gradations which we already know, and which doubtless exist in far greater number, is no great or impossible step; and the truth of the old Linnæan maxim, *Natura non facit saltum* (Nature takes no leaps), is here again proved in the most striking manner in favor of the theory of evolution. As the animal in common with the whole organised world is physically developed, transformed and improved, so it is also developed psychically to that height whereto the nature of its organisation and life conditions enable it to reach: it is only the extreme shortness of our experience which does not distinctly prove this to us, and allows us to think that all is standing still, just as on account of its vast distance the heaven of the fixed stars is to us a picture of an eternal remaining rest, while, in reality, all is movement and change there. So can the intellectual and physical life of animals now only be rightly understood to-day by those who seek and find the key of the knowledge of its existence in the present in the many millions of years reckoned as its past, and thereby utilise as ladders the vast mass of transitions and gradations present also to-day. He who does not know or does not understand this key for the solution of the great enigma, he stands—to put it somewhat drastically—with regard to this question, as the ox on the mountain, and can only arrive at the absurd opinion, despising all facts, which was maintained, for example, by Professor John Huber, of München, in his articles on scientific questions of the day (Supplement of the *Allgemeinen Zeitung* of July 14th, 1874), where he alleged that animals make no progress, that they discover nothing, that they build their nests as at first, that they indeed gain experience, but cannot impart their know—

ledge to their companions and their successors, and cannot utilise it themselves; or by Professor Carus, sen., who rose in his "Comparative Psychology" (1866, p. 191) to the astounding assertion: "The spider's web is woven by the spider involuntarily, and involuntarily are the insects caught therein used as food." Has Herr Carus, who considers himself capable of writing a Comparative Pyschology or Mental Science, never heard that spiders, like almost all animals, use very admirable discrimination in the choice of their food, although, also like all other animals and like man, they are subject to occasional error? Mr. Moggridge, on one occasion of watching the trap-door spiders at night, brought an accidentally-captured beetle (*Chrysomela Banksii*) into the immediate neighborhood of a half-open trap-door, out of the crack of which lay the fore-legs of the inhabitant. The door immediately flew open, the spider darted on the beetle, and drew it into its house, whereupon the door shut close. But after a few seconds, remarkable to say, the door suddenly opened again, and the beetle came alive and unhurt into sight—*i.e.*, the spider threw him out again. Clearly, the beetle, which was far too small and weak to make any important resistance, had either some peculiarity which made it unpleasant to the spider, or was unfitted for food, and was therefore let go uninjured. A few minutes afterwards Moggridge brought a woodlouse (*Oniscus*) to the door of the spider, and saw this pulled in, and *not* let out again. What becomes, in such cases, of the Carusian theory of the involuntarily eaten spider's food?

But the *ne plus ultra* of this melancholy kind of wisdom is manifested by Professor Fr. Körner ("Instinct and Free Will," 1874), who starts from the opinion that "all animals of the same species do all the same things and in the same way since thousands of years ago," and explains the spider's web in the following intellectual fashion: "The spider must make a web, because the stored up spinning material in its body hurts it." A man, indeed, like Herr Körner—who answers the question, asked by himself, why the swallow should build a mud nest, by saying "Because it is a house-swallow;" or gives as explanation of the beaver-dam that "the beaver must build thus because it is in some measure in his feet;" or maintains as to the dog that he "knows nothing of duty and conscience, of truth and devotion;" or has nothing better to say of a human child than that it is "only a wax-like piece of flesh;"—such a man ought not to expect anyone to trouble himself seriously with contradicting his printed folly.

He who objects to the application of the evolution theory to the above-mentioned facts on the ground that it is not clear why, in spite of such long past ages, many imperfect and·as it were primitive forms of the nests of the earth-working spiders should exist side by side with the more perfect, forgets that it is not different with us men, although we are wont to claim progress as a human privilege, and although the antiquity of the human race on the earth must be reckoned in all probability as of hundreds of thousands of years: not only is the number of cave-dwellings and huts of the most primitive kind in which men live, far in excess of that of the houses and palaces of civilised nations; but also the comparative distance between them appears far more striking than that which we have found between the abodes of trap-door spiders.

Finally, before leaving the interesting spider-folk, we may remember a trap-door spider discovered by Dr. Livingstone, the famous African traveller, near the Delilo Lake in South Africa. "A large reddish spider (*Mygale*,)" says the discoverer, ("Pop. Accounts of Travels in South Africa," ch. xvii., p. 221,) "which the natives call Sclüli, is seen running about here with great nimbleness. Its nest is closed in the most ingenious manner with a lid or door, about the size of a shilling, hanging from a hinge. The inner surface of this door is covered with a pure white silken paper-like material, while the outside looks exactly like the surrounding surface of the ground, so that when the door is shut it is impossible to find the nest. The hole can therefore only be seen when the inhabitant has gone out and has left the door open behind it."

THE BEETLES.

CHAPTER XXVIII.

BEETLE INTELLECT.

FROM the spiders, which depart so far in their bodily organisation from the insect type proper, that they are set aside in a special class of the articulate animals under the name of *Arachnida*, let us turn back to a group which represents this type in its highest perfection or development—the universally known, and almost inconceivably various in form, beetles or *Coleopteræ*. If only the representatives of different species in the countless collections of this specially beloved object of collectors were reckoned up, they would mount to far over a hundred thousand; and were we to judge by the interest which these creatures exercise over so many people we should have to expect something very special from them for the object of this work. But in reality, both as to their mechanical ability and as to their other intellectual capacities, they stand far, far behind the hitherto studied representatives of the articulate animals, although a more exact observation of their habits and action than we as yet possess would here also, without doubt, bring to light in some species and genera a perhaps hitherto not suspected intellectual capacity. Debey maintains ("*Beiträge zur Lebens und Entwicklungs Geschichte der Rüsselkäfer aus der Zunft der Attelabiden,*" Bonn, 1846, and quoted by Perty, *loc. cit.*, p. 300,) that the so-called funnel-spinner, *Rhynchites betulæ*, possesses the highest instinct among beetles, and that it with other *Attelabides* should be placed near, if not above, the honey-bee and the ant. In general, however, such a judgment is doubtless inapplicable to the great order of the beetles, or beetle-like insects. Their helplessness, their clumsy ways, their manner of flight and work, their want of ordered social life, and similar facts, point to the lower intellectual grade which they take among their allies, and they may be described as among insects what porters are among men. But as individuals are found among porters whose intelligence raises them far above the crowd of their colleagues, so among beetles are found some proofs of a very far-reaching reasoning or reflective power. The most well-known by its habits is the burying-beetle (*Necrophorus*), which also manifests a small commencement of social life, in that several of them unite together to bury under the ground as food and shelter for their young some dead animal, such as a mouse, a toad, a mole, a bird, etc. The burial is performed because the corpse, if left above-ground, would either dry up, or grow rotten, or be eaten by other animals. In all these cases the young would perish, whereas the dead body lying in the earth and withdrawn from the outer air, lasts very well. The burying-beetles go to work in a very well-considered fashion, for they scrape away the earth lying under the body so that it sinks of itself deeper and deeper. When it is deep enough down, it is covered over from above. If the situation is stony, the beetles with united forces and great efforts drag the corpse to some place more suitable for burying. They work so diligently that a mouse, for instance, is buried within three hours. But they often work on for days, so as to bury the body as deeply as possible. From large carcases, such as those of horses, sheep, etc., they only bury pieces as large as they can manage.

Gleditsch, to whom botany and farming owe so much, put four *Necrophori*, with their young, into a glass vessel filled with earth, and found that these insects, in a space of fifty days, buried no less than four frogs, three small birds, two grasshoppers, and a mole, as well as the entrails of a fish and two pieces of ox-lung. He saw a single beetle, which he left without help, bury a dead body in the earth by unceasing efforts and great sense or skill.

An observer, to try the acuteness of this creature, tied a dead mouse on a cross of wood, which would hold the mouse up after the burrowing beneath it. But when the beetles saw that after they had burrowed, the mouse did not fall into the hole, they undermined the wood as well, until the body fell in.

They went yet more sensibly to work when a friend of Gleditsch fastened a dead toad, which he wanted to dry, on a stick, and put the stick in the ground. The beetles were attracted by the smell, at once recognised the difficulty of the situation, and undermined the little post until it fell over, whereupon they buried toad and stick. Dead animals hung on a stick by means of a thread, so that they rest on the ground but cannot sink in, are obtained by the *Necrophori* by undermining the hindering stick.

The *Necrophori*, like the greater number of their beetle allies, possess a very well-developed stridulating, or rasping apparatus, by means of which they emit a very broken, jarring note, which, perhaps—apart from other purposes— serves to call them together for the performance of their common work. They can also, like all insects, understand and communicate with each other by means of their antennæ. The same is, of course, true of beetles without exception, and there can be no doubt that they use their often very various and even curiously made antennæ, just as do bees and ants, for mutual comprehension, even though the communications which they have to make to each other are of a far simpler nature than with the animals named. Herr George Goelitz writes as follows to the author from Marysville, Marshall County, Kansas (North America), under date Dec. 25, 1875 :—" Last summer, in the month of July, I was one day in my field, and found there a mound of fresh earth like a molehill, on which a striped black and red beetle, with long legs, and about the size of a hornet, was busy taking away the earth from a hole that led like a pit into the mound, and levelling the place. After I had watched this beetle for some time, I noticed a second beetle of the same kind, which brought a little lump of earth from the interior to the opening of the hole, and then disappeared again in the mound; every four or five minutes a pellet came out of the hole, and was carried away by the first-named beetle. After I had watched these proceedings for about half-an-hour, the beetle which had been working underground came out and ran to its comrade. Both put their heads together, and clearly held a conversation, for immediately afterwards they changed work. The one which had been working outside went into the mound, the other took the outside labor, and all went on vigorously. I watched the affair still for a little longer, and went away with the notion that these insects could understand each other just like men." Klingelhöffer, of Darmstadt (in Brehm, *loc. cit.*, ix., p. 86), says:—" A golden running beetle came to a cockchafer lying on its back in the garden, intending to eat it, but was unable to master it; it ran to the next bush, and returned with a friend, whereupon the two overpowered the cockchafer, and pulled it off to their hiding-place."

Many other beetles, in addition to those named, have been observed to summon each other for assistance. This has been seen most strikingly exemplified in the famous Scarabæus, or sacred beetle of Egypt (*Ateuchus*, or *Scarabæus sacer*), whose remarkable method of propagation seemed so marvellous to the ancients that they dedicated it to the sun, and that the old Egyptians paid it divine honors, and placed it in their temples, hewing it in colossal size out of stone. The wiser Romans, on the other hand, narrowed down the worship to carrying beetles carved in stone as amulets. Sometimes real beetles were mummified and kept in special vessels. The *Ateuchus* has the remarkable habit of making pellets of dung, one or two inches big, in which it places its future young, and then rolls them in front of it until they are round and firm, and have arrived at the spot where it wishes to bury them. This turning and rolling of its egg-enclosing dung-pellets was taken by the ancients as a symbol of the movement of the earth, and hence their reverence! Cow-dung is preferred by the *Ateuchus* for its object before any other, but, in default of this, it will use that of sheep or goats. The work itself is very carefully performed, and at every couple of steps it halts, to see if the pellet is firm enough. In order to find the place where it may most suitably be buried, the *Ateuchus* has often a long way to go, in which case one mate generally pulls while the other shoves or pushes. If it finds any roughness of ground on the way, it lifts up the pellet, which sometimes attains the size of a small apple, on its broad, strong head, as with a lever. Sometimes it chances that the pellet falls into a hole, or some unevenness in the ground, where the beetle does not want it, and out of which it is unable to lift it either alone or with the help of its mate. The beetle then suddenly leaves its ball, spreads its wings, and rises into the air. If the observer is patient enough to see the matter out, he will see the fugitive return after awhile, accompanied by two, three, four, or five comrades, which now set to work together, and set the pellet

rolling again. It thus happens that, on stony ground, several beetles are not seldom seen busied with a single ball. At last, arrived at a suitable spot, a hole is dug in the ground with the strong, toothed fore-feet, which can work like a digger's spade, the pellet is dropped in and the earth shovelled over.

According to M. P. de la Brûlerie (A. Murray, " Journ. of Travels," vol. I., ˉ1868) the male *Ateuchus* uses his jarring rasp or stridulating note to encourage the female when making the dung-pellet for the future young, as well as from uneasiness when she is away.

The *Scarabæi* are specially eager after freshly fallen dung, because, mixed with earth, it best serves their purpose. As a rule they leave the pellet to dry a little in the sun before they begin to roll it.

The *Oncideres amputator*, a species of house-beetle (*Lamia*) living in the tropics, deserves notice, which gnaws round the bark and alburnum of young twigs, and thereby causes their death or fall. M. Foullet, director of the hothouses of the Natural History Museum in Paris, was in a house near Rio Janeiro and heard every night the rustle of the falling twigs of a tree, the *Acacia Lebbeck*. These twigs seemed to be sawn off all round, and as only their inner part remained they fell either of their own accord or in consequence of the wind. On whom was the guilt to be laid? Without doubt it was the negroes of the property who wanted to play their master a trick. But˙ the traveller soon noticed that a *Lamia* was often on the cut twig, and that it must be the evil doer. One of the twigs was investigated, and was found filled with the living larvæ and pupæ of the *Oncideres*. The object of the destroyer was now clear; it wanted, by cutting through the alburnum, to prevent its larvæ from being drowned or injured by an overflow of sap!

A not less refined care for the bringing up of its brood is shown by many of the already named weevils, or *Rhynchophora*. The *Rhynchites auratus*, belonging to the group of so-called fruit-piercers, seeks the sunny side of the apple, raises a bit of the peel, lays an egg in the little hole it hollows out, and then lays down the peel again so carefully that the place can scarcely be seen. The hatched larva does not live entirely on the pulp of the apple, but bores through it, gets into the core, eats the pips, works right through the apple, lets itself fall and turns into a pupa in the ground. The *Rhynchites betulæ*, described by Debey, cuts the leaves of the birch in a very artistic manner from the edge inwards, so as to be able to roll up the edges and make a funnel in which to lay its egg. As it partly gnaws through the middle bundle of the leaf at the same time, the latter gradually dries up, and when the larva is hatched it finds food already prepared for it in the dry substance of the leaf. When the withered leaf at last falls off it turns into a pupa in the ground. The *Rhynchites betuleti*, which does great mischief in the vineyards of the Rheiu and the Moselle, rolls up several of the top leaves of a twig in the most ingenious way—after it has pierced or notched either the stem or themselves, so as to have caused the partial withering of the leaves, making them pliable—into the form of a cigar, smearing their edges with a sticky secretion, and smoothing the roll with its abdomen. According to Nördlinger's observations the little artist performs his difficult task with wonderful reflexion and appreciation of each special situation, as well as with astonishing skill, strength, and patience. Further, among the Attelabi, only the females possess this thoughtful industry, inherited from generation to generation, and the intellectual superiority of the female sex over the male stands out again here, as among so many other, and indeed most, insects.

The *Attelabus curculionoides* treats the leaves of the oak, and the *Apoderus coryli* those of the hazel, just as the *Rhynchites betulæ* does those of the birch.

Very quarrelsome beetles are the allied and predacious *Cicindelæ*, belonging to the family of the running-beetles, which Linnæus has justly named the tigers among insects (*Tigrides insectorum*). They are all strong and swift robbers, hunting living insects and falling with tigerish blood-thirstiness upon their victims, striking them with their sharp-pointed upper jaws, tearing them in pieces, and swallowing them piece by piece. In spring the graceful active *Cicindela campestris* or field *Cicindela* may be seen in the sunshine on all sandy places, with its emerald green wings and copper-red spots glittering like fire in the sun. Scarcely an insect can withstand its attack, while it has itself little or nothing to fear with its hard body covering. The larvæ of the *Cicindelæ* have all the voracity of their parents, and they manage to satisfy it in a very ingenious fashion without danger to their soft bodies. They dig chimney-shaped pits in the ground with their strong pointed-feet, in and out of which they climb by alternate motions of their bodies, like a sweep in a chimney. Holding their heads quietly at the entrance of their hole and to a certain extent stopping it up thereby, they wait with tireless patience for insects running over. As soon as one of these treads on the dangerous spot the larva pulls its head back quickly and the insect falls into the hole and becomes the prey of the robber. This game is continually repeated. When the larva is ready to spin its cocoon, it simply walls in the entrance of its hole.

If the field *Cicindela* is put with other insects which serve it as food, as house-flies, worms, caterpillars, other beetles, etc., it tries first of all to sever the head from the body of its victim, or to tear and bite off wings and legs, so that it cannot escape. It then cuts from within outwards, so that only the empty shell remains. The murderer often tears open the abdomen of its prey, and devours the entrails while the latter is still alive and trying to escape. Caterpillars, worms, or soft larvæ are also gnawed from the abdomen upwards, and slowly eaten up from behind forwards while still alive.

The predacious and greedy *Staphylinus* species also, which show much resemblance in their life-habits with the running-beetles, act in the same way as the *Cicindelæ*. Dr. Nagel, of Schmölle, watched the fight of *Staphylinus maxillosus* with the larva of a *Tenebrio molitor*, the so-called meal-worm. At first the grasp of the beetle slid off from the hard, smooth, chitinous somites of the worm. At last, however, it caught the worm so tightly by the neck that it could not get free, spite of all twistings and turnings. They both wrestled together like two dogs, which seize each other by the forefeet and rise on their hind-legs. At last they fell, and the head of the worm was torn from the convulsively twisting body. The beetle now devoured the worm, with great anatomical skill twisting off one somite after another from the soft abdominal wall, so as to be able to get at the inside. In other cases the beetle at once gnawed at the abdominal wall, and began thence its work of destruction.

An artificially-induced fight between a *Cicindela* and a running-beetle (*Carabus*) had no result, both creatures seeming afraid, and letting go as soon as they seized each other. More severe are the battles waged between the rival beetles for the possession of a female, which Darwin ("Descent of Man," p. 299) has described in detail.

Nagel put four *Feroniæ* (a species of common running-beetle) with a *Staphylinus niger*. The *Staphylinus* seized one of them between head and thorax, and a furious struggle took place, the three others looking on quietly for awhile. At last, however, they ran to the wrestlers and attempted intervention, but speedily took to flight when they received a push from the latter. At last one of the *Feroniæ* took courage, and bit the abdomen of the *Staphylinus*. It, however, did not allow itself to be diverted, but finished its murderous work. This so terrified the three other beetles that they promptly crept into the ground. Could the intellectual struggle between fear of an overwhelming foe and the desire to help a comrade in distress have happened otherwise among men?

A striking example of the sharpness of a beetle is related by G. Berkeley ("Life and Recollections," vol. II., p. 356), without the species of beetle being given. While out walking he saw a beetle pulling along something heavy. Kneeling down, he observed that it was the body of a large, brown, apparently dead spider, and that the beetle was more than half-an-inch long, resembling a large fly in appearance, of a dark color, with shining yellow or red sides. The observer took up the beetle on the end of his stick, to look at it more closely, but it fell down off the stick and lost its prey, or let it fall. Both reached the ground about a foot apart. But the beetle at once began to look for its prize, apparently being guided by its scent. When it again found the spider, it approached it carefully, avoiding its head, and touched its side with one of its forefeet, so as to assure itself that the spider was really dead. On feeling quite convinced, it seized its prey again and ran on. A few steps further it laid it down, however, and went on without it. The observer followed on his hands and knees, and saw it leave the sandy path and climb up on some heather. It here rested for a few minutes, whereupon it carried the spider up to the top of the heather and hung it up between the twigs, of which it had clearly known beforehand that they would do to take and keep its game.

When this was done, it went down again, and disappeared between the roots of the heather, as though seeking new prey. The observer now examined the twig, and convinced himself that the beetle had not only chosen the most suitable twig for hanging up its food, but had also managed the suspension in the most suitable way. For a sharp shake of the twig did not make the spider fall down. "Who will now," adds Mr. Berkeley, "deny to the brain of the insect the power of reflexion or of judgment? The beetle thought that if it did not hang up its prey, it might fall into the hands of other hunters, and therefore took all possible pains to find out the best store-room for it."

THE ANT-LION.

FINALLY, the famous ant-lion must be remembered (*Myrmecoleon*), which, although not belonging to the beetles, has always attracted the attention of lovers of nature by its remarkable way of hunting. It is the larva of an insect belonging to the order *Neuroptera*, the common *Myrmeleon* (*Myrmeleon formicarium*), and is therefore nearly allied to the Termites. It obtains its food in the following ingenious way, which vividly reminds us of the manners of loan-office vampires among men. For both live on harmless unsuspecting victims, whom chance throws into their jaws by the help of a trap set in most unsuspected and therefore most dangerous fashion. In dry sandy places the ant-lion makes its well-known funnel-shaped pits, first making a circle in the sand which marks the circumference of its hole, and then makes the hole itself by shovelling out the sand. This is done in the following fashion: it loads the sand on its flat head, using one fore-foot as a shovel, and then jerks it so forcibly out of the circle that it flies several inches away. During this process it always goes backwards, until it comes again to the place whence it started. It then makes a new circle, and hollows out a new pit in similar fashion, until it gradually comes to the centre of the funnel. But in order not to tire out one leg too much with continual work, it draws alternate circles in opposite directions and uses the other foot. If it comes across little stones or too coarse grains of sand, it throws them over the edge of the hole. But if the stone is too large to treat thus, it gets it on its back very cleverly, climbs slowly and carefully up the side and throws it outside the pit. It is sometimes seen thus to move stones four times as heavy as itself. If the stone be round, this proceeding is complicated by no light difficulties, and it often rolls down again. But it patiently picks it up again, and tries to carry it away, sensibly using the path made by the fallen stone. If the attempt fails, however, time after time, the creature gives up its Sisyphus-task, and makes another funnel.

When the pit is complete, the sly vampire hides itself in the sand at the bottom, so that none of it is to be seen, and only the open points of its long strong jaws emerge. As soon as a little insect comes to the edge of the pit, the soft sand shifts under its feet and it falls in, to be seized in the jaws of the hunter. The victim is sucked until nothing remains but its husk, which is then flung over the edge of the hole. Among men the sucked-out husk is often left alive, but not seldom shoots itself to death.

The unfortunate insect, as soon as it feels the ground giving way under its feet, naturally tries to escape by climbing up the sides of the pit. But the ant-lion knows how to make these efforts useless, by throwing a rain of sand over the creature with its broad head, which catches it in rolling down again and brings it within reach of the robber. It is generally ants which fall victims to it in this way. Hence its name.

When all is over, the robber comes out from its hiding-place, and sets in order again any places that may have been injured, and then lurks in wait for fresh prey. Sometimes it is deceived, and lets itself be induced to throw up its sand-rain as often as is wished, if by aid of a stalk or little stick some tiny lumps of sand are pushed down and let to fall into the centre of its funnel. Human ant-lions, too, are sometimes tricked!

THE END.

POSTSCRIPT.

THERE are few points in the foregoing book as to which any words of mine are needed. The differences between the original and the translation are very slight. I have omitted a few lines from the preface to the third edition, referring merely to changes which do not touch the English reader. I have also divided the book into Parts and Chapters instead of printing it continuously, as in the German, from beginning to end without break. I have in a few cases added the date of the edition of the English books quoted, and have corrected the page reference when that of the edition used was not the same as that quoted from by Dr. Büchner. In most cases of quotations from English books, I have taken them directly from the originals, instead of retranslating. Here and there I have added a note, marking it by placing it between square brackets and signing it "Tr." A few points touching which I was doubtful have been referred to Dr. Büchner, so as to insure perfect accuracy, and my thanks are due to the great German author for his ready and over-kindly assistance. I also owe thanks and recognition to my friend Dr. Edward Aveling for reading over a great part of the proofs.

ANNIE BESANT.

January 13th, 1881.

INDEX.

INTRODUCTION, 1—35.

CHAPTER I.

PAGES,

HISTORICAL REVIEW ... ... 1—10

Opinions on "instinct" of Anaxagoras, Socrates, Plato, Aristotle, Pliny, Virgil, Plutarch, Galen, Celsus, Descartes, Cæsalpinius, Leibnitz, Bajer, Reclam, Condillac, Buffon, Bonnet, Bonjeant, La Fontaine, Leroy, Kant, Herder, Agassiz, Huxley—Voice Organs of Man and Animals.

CHAPTER II.

INTELLIGENCE AND "INSTINCT" ... ... 11—22

Instances of Intelligence—Blunders of Animals—Changes of Food " Instinct "—Changes of Building " Instinct " in Beavers, Sables, Birds, Bees, Oriole, Bittern, Palmswift, Puffin.

CHAPTER III.

" INSTINCT " (continued)... ... ... 23—35

" Instinct " of Chickens, Ducklings, Goslings, Turtles—Smell of Insects—Dams of Beavers—" Instinct " in Man—Animal Psychology.

ANTS AND ANT LIFE, 39—184.

CHAPTER IV.

GENERAL CHARACTERISTICS ... ... 39—53

Place of Ants in 'Nature—Resemblance between Ants and Men—Variety of Character—Courage—Size and Construction of Brain—Effect of Injuries to Brain—Use of Antennæ—Love of Cleanliness—Anatomy.

CHAPTER V.

ANTS IN HISTORY ... ... 54—70

Arabia—Jewish Scriptures—Mishna—Hesiod—Horace—Virgil—Plautus—Pliny—Ælianus—Aldrovandus—La Fontaine—Plutarch—The Ant Republic—Inferiority of Males and Fertile Females—Wedding Flight—Queens—Care of Young—Larvæ—Nymphæ—Young Ants—Their Education—Equality of Ants—Working Queens.

CHAPTER VI.

ANT ARCHITECTURE ... ... 71—89

The Ant-heap—Huber's Description of a Nest—Method of Building—Variety and Adaptation to Circumstances—Nests in Trees—Nests of Harvesting Ants—Intelligence under Difficulties—Doors, Sentries, and Methods of Defence of Nests—Changes of Dwelling—Ants in New South Wales.

CHAPTER VII.

ROAD-MAKING ... ..., 90—101

Covered Ways—Dépots—Colonies of many Nests—Ant-town in Pennsylvania—Working at Night—Marching in Africa—Sa-ubas in South America—Umbrella Ants in Central America—Cutting Ants—Tunnels of Sa-ubas—Plundering Stores.

CHAPTER VIII.

HARVESTING ANTS ... ... 102—115

Gathering Provisions—Rubbish Heaps—Preventing Germination of Corn—Plundering—Mistakes—New Indian Ants—Mexican Agricultural Ants.

CHAPTER IX.

CATTLE AND MILKING ... ... 116—127

Aphides: Shutting them up—Milking them—Crossing birdlime, etc., to get at them—Forming a colony—*Myrmecophila*—Honey Ants.

CHAPTER X.

INTELLIGENCE AND LANGUAGE... ... ... 128—135.

Intelligence in reaching Stores of Food—Fine Scent—Communication by Antennæ—Ants in Omotepe with Dead Birds—In a Monastery with Sugar.

CHAPTER XI.

SLAVERY ... ... 136—152

Mildness of Slavery among Ants—Amazons and their Slaves—Slave Hunts—Attacks on Nests of Slave Species—Carrying off of Slave Pupæ—Civil War.

CHAPTER XII.

SLAVERY (continued) ... ... 153—163

Sanguine Ants—Intelligence in Battle—Slave Hunts—Besieging a Nest—Slave Species—Care of Sick—Gymnastic Sports.

CHAPTER XIII.

FRIENDSHIP AND ENMITIES ... ... 164—169

Mutual Feeding—Experiments with Flasks—Hostile Ants meeting—Wounded and Dead—Meeting after separation.

CHAPTER XIV.

WARS AND BATTLES ... ... 170—184

Battles between different Species—Between the same Species—Alliances—Various Methods of Fighting—*Pheidoles*—Soldiers—*Eciton* or Driver Ant—Contrast between Workers and Soldiers.

THE TERMITES OR WHITE ANTS, 185—207.

CHAPTER XV.

TERMITES AT HOME ... ... 187—198

Large Buildings—Interior of Nest—Defence by the Soldiers—Repairs by the Workers—Covered Roads—Soldiers and Workers—Queen—Wedding Flight.

CHAPTER XVI.

TERMITES ABROAD ... ... 199—207

Destructiveness—Introduction into Europe—Bates's Account of their Life and Ways.

THE BEE NATION, 209—289.

CHAPTER XVII.

ROYALTY ... ... 211—227

The Queen Bee—Drones—Massacre of the Drones—Female Supremacy — Struggles between Queens — Emergence of Young Queens—Development of Royal Larvæ—Loss of Queen—Introduction of a Stranger Queen.

CHAPTER XVIII.

SWARMING ... ... 228—241

Signs of Swarming—Swarming—Artificial and Natural — Settlement in a New Home — Killing of Young Queens by Old—Wedding Flight—Return of the Queen—Egg-laying—Determination of Nature of Eggs—Abdication of Queen.

CHAPTER XIX.

DOMESTIC WORK ... ... 242—254

Work of Younger Bees—Feeding of Larvæ—Preparation of Honey and Storage thereof—Pollen—Cleaning the Hive—Rebuilding the Cells—Change of Building to Suit Circumstances—Propolis—Cleanliness of Bees—Burial of Dead—Ventilating Bees—Details on Cell Building.

CHAPTER XX.

ACTIVITY ABROAD ... ... 255—264

Gathering Nectar and Pollen—Guards—Enemies—Language—Scent—Memory—Virgil and Shakspere on Bees.

CHAPTER XXI.

MONARCHY, SOCIALISM AND INSTINCT ... ... 265—272

Constitutional Monarchy—Communism—" Instinct " of Work—Robber Bees—Bees with Sugar Refineries—Living on Humble Bees—Artificial Feeding.

CHAPTER XXII.

CELL-BUILDING ... ... 273—283

Shape of Cell — Gradations of Shape — Meliponæ — Hexagons—Long Development—Queens and Workers—Wars—Greeks and Bees.

CHAPTER XXIII.

OTHER SPECIES OF BEES ... ... 284—289
Osmia, or Mason Bee—Melipona—South American Bees—Tapestry Bee—Rose Bee—Carpenter Bee—Humble Bee.

THE WASPS, 291—309.

CHAPTER XXIV.

GENERAL DETAILS ... ... 293—301
Fertile Females—Workers—Nests—Hornets' Nests—Intelligence—Wasp and Spider—Irritability.

CHAPTER XXV.

VARIOUS SPECIES OF WASPS ... ... 302—309
Polistes—American Wasps—Wall or Mason Wasp—Sand Wasp—Butcher Wasp—Bee Eater—Ichneumonidæ.

THE SPIDERS, 311—338.

CHAPTER XXVI.

GENERAL DETAILS ... ... 313—321
Dislike of Spiders—Spiders' Webs—Stretching of Webs—Cleanliness—Love of Music.

CHAPTER XXVII.

VARIOUS SPECIES OF SPIDERS ... ... 322—338
Water Spider—Hunting Spider—Bird-catching Spider—Trap-door Spider—Its Nest—Moggridge's Investigations—Various Kinds of Nest—Australian Trap-door Spiders—Transitional Forms of Nest—Evolution and its Opponents.

THE BEETLES, 339—351.

CHAPTER XXVIII.

BEETLE INTELLECT ... ... 341—351
The Burying Beetle and its Cleverness—Power of Communication — Ateuchus — Oncideres Amputator — Rhynchophora—Cicindela—Staphylinus—A Beetle's Larder—The Ant Lion.

www.ingramcontent.com/pod-product-compliance
Lightning Source LLC
Chambersburg PA
CBHW020310240426
43673CB00039B/759